SpringerBriefs in Energy

For further volumes:
http://www.springer.com/series/8903

Mrinmoy Majumder · Soumya Ghosh

Decision Making Algorithms for Hydro-Power Plant Location

 Springer

Mrinmoy Majumder
National Institute of Technology
Agartala
India

Soumya Ghosh
Greater Kolkata College of Engineering
 and Management
Kolkata
India

ISSN 2191-5520   ISSN 2191-5539 (electronic)
ISBN 978-981-4451-62-8   ISBN 978-981-4451-63-5   (eBook)
DOI 10.1007/978-981-4451-63-5
Springer Singapore Heidelberg New York Dordrecht London

Library of Congress Control Number: 2013936523

Printed on acid-free paper

Springer is part of Springer Science+Business Media (www.springer.com)

Contents

Chapter 1
Introduction

Abstract The global energy scenario highlights the growth in demand for fossil fuels due to the continuous development in technology and the socioeconomic scenario of the world. This increase in demand of fossil fuels and utilization of the same to sustain development has enforced stress on the resources. Also, uncontrolled use of fossil fuels has increased the total concentration of greenhouse gases, which in turn has become a major cause of global warming. As a result the climate of many places has displayed abnormality. The pollution content of various places is also aggravated due to the rampant use of fossil fuels. That is why an alternative to fossil fuel is now being searched. Some of the sources of energy like solar, wind, and hydro-energy are infinitely available but the technology to convert it into utilizable form is expensive. Among all these renewable energy sources, hydro-energy is found to be relatively inexpensive and available at a greater phase of time than other similar kinds of energy resources. However,most of the factors which are considered in the feasibility studies for HPP is location dependent. The requirement of population displacement from the project watershed depends on location of the project. The amount of utilizable hydro-kinetic energy also is a function of both space and time. Not only the location dependency of factors but influence of all the factors on generation capacity is not uniform. For example the influence of amount of flow available in the project location is more important than the area of forest which are required to be removed from the project area. That is why both the importance of the factors and its location dependency must be considered in any feasibility studies for HPP. If the site selection is performed logically and scientifically, considering the importance of all the socioeconomic, geophysical, and logistical factors, then only such project may optimally satisfy the present demand for energy. In this regard decision-making algorithms like neural network, fuzzy logic, bat algorithms, and analytical hierarchy process along

M. Majumder and S. Ghosh, *Decision Making Algorithms for Hydro-Power Plant Location*, SpringerBriefs in Energy, DOI: 10.1007/978-981-4451-63-5_1, © The Author(s) 2013

with hybrid models like neurogenetic and neuro-fuzzy which are popular for their intuistic decision making abilities were applied to identify the ideal sites for installation of hydropower.

Rapid increase in urbanization, ever-growing population along with global warming-induced climate change and large-scale extraction of natural resources has in recent times disturbed the natural equilibrium that provides the requisite resources for sustenance of mankind.

Along with the environmental degradation the demand for fossil fuels has raised manifold in the last ten decades due to the economic and technological developments to satisfy the luxury needed by the population. As a result, the pressure on the conventional fuel sources, which are finite, has also increased.

That is why engineers and scientists are looking for alternative sources of energy to satisfy this increase in demand. Solar, wind, and hydropower were found to be some of the major sources that can be suitable alternatives for mitigating energy scarcity observed in many places around the world.

There are many other sources of energy that can be utilized as alternatives to the conventional energy depositories. Among them, with respect to price, availability, and regularity, solar, wind, and hydropower were found to be the potential alternatives for replacing the utility of fossil fuels.

1.1 Global Energy Scenario

Figure 1.1 depicts the contribution of different types of energy sources from 1990 to 2008 (Eenergiläget in Sweden 2011). The figure shows that the contribution from fossil fuel is still the highest even in the year 2008, but the utility of Renewable Energy has also increased from 1,308,000 to 1,849,000 TWh. At the same time the use of nuclear energy is also aggravated from 611,000 to 82,000 TWh.

The regional energy use in the said time frame has grown by 170, 146, 91, 70, 66, 20, and 7 % respectively in the Middle East, India, Africa, Latin America, USA, and the EU-27 block (IEA 2012).

Globally, energy use was found to increase by 39 %. While the global population has increased by 27 % from 1990 to 2008, the energy use per person has increased by 10 %. The dependency on fossil fuel is also found to be shifting toward renewable sources of energy (Fig. 1.1).

1.2 Threats to Natural Resources

The crisis for conventional energy sources is not the only reason why experts are enforcing governments to shift toward renewable archives of energy. The pollution exerted from using fossil fuels is one of the major concern for adaptation of

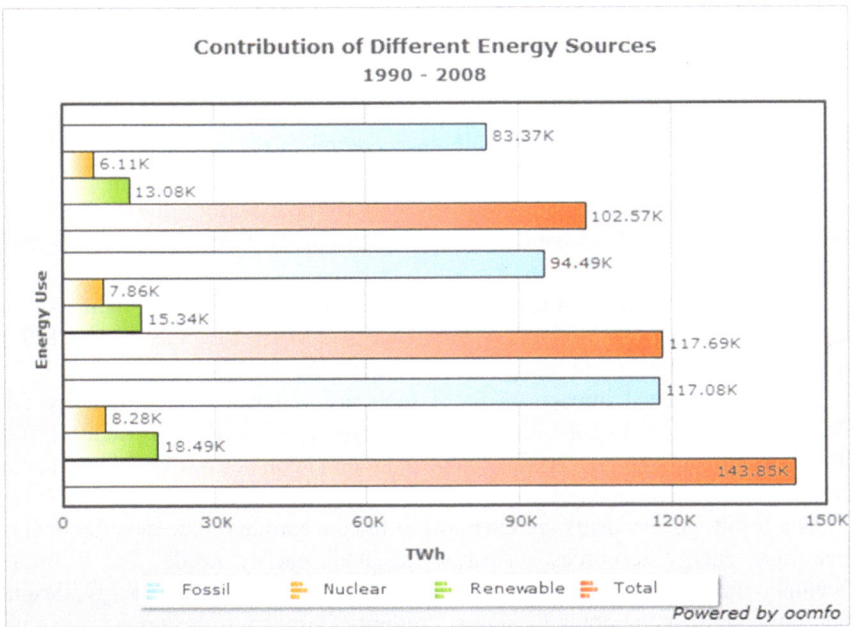

Fig. 1.1 The contribution of different energy sources to satisfy the energy demand from the global population

different kinds of energy sources which are more environment friendly and may not influence the global pollution scenario.

The burning of fossil fuels contributes to global warming, the pollution of the air, water, and land, and is one of the major factors for production of acid rain. During the combustion of fossil fuels huge amounts of carbon are released into the air which again increases the concentration of greenhouse gases in the atmosphere. The result of this entrapment results in global abnormality in the climatic trend.

The temperature rise due to global warming contributed to the melting of the ice cap in the polar region, which has decreased the salinity of the ocean endangering many organisms that are dependent upon a certain level of salt concentration. Sea level ascent also poses a serious risk to many cities and settlements located close to the sea level which can entirely disappear under water.

Besides sea level rise, climatic abnormalities due to global warming will result in altered habitats, extreme weather conditions, drop in crop yields etc. In countries like Africa, the Middle East, and India, water availability for irrigation and drinking will be less predictable due to large-scale variations in the rainfall pattern. Droughts will be more frequent, creating pressure on agricultural production and diversity.

Up to three billion people could face increased water shortages by 2080. Air and water pollution resulting from the extraction and use of fossil fuels can lead to significant health and environmental degradation.

The economic consequences are also significant. The rise in economic development all over the world, especially in the BRICS countries, and the demand for raw materials and energy to sustain their developments is causing pressure on the fuel reserves. Due to this, uncontrolled extraction of the resources to supply to this demand has become scarce enforcing an accretion in the prices of the products (TAL 2012).

The ill effects of using fossil fuels are also described in various literatures such as Sovacool (2009), Goldemberg (2007), Lohmann (2006), Jaccard (2006), Vitousek et al. (1997), etc.

The threat to water resources from the combustion of fossil fuels is also depicted in Hanlon et al. (2013), Bradley et al. (2006), Pimentel et al. (1997), Frederick et al. (1997) and many other related literatures.

The environmental impacts of fossil fuels are delineated in: Mukherjee and Chakraborty (2013), Jaccard (2006), Laurance et al. (2001), Beckerman (1992), Davies (1990) and many other literatures and reports published from governmental as well as non-governmental organizations.

As a result, governments are encouraged to concentrate on the development of renewable energy resources for satisfying their energy needs. For example, Denmark and Germany have started to make investments in solar energy, despite their unfavorable geographic locations. Presently Germany is the largest consumer of photovoltaic cells in the world.

Denmark and Germany have installed 3 and 17 GW of wind power respectively. In 2005, wind generated 18.5 % of all the electricity in Denmark. Brazil invests in ethanol production from sugarcane, which is now a significant part of the transportation fuel in that country. Switzerland is planning to cut its energy consumption by more than half to become a 2,000 W society by 2050 and the United Kingdom is working toward a zero energy building standard for all new housing by 2016.

1.3 Potential of Hydropower as Possible Alternative

"As one of the earliest renewable energy resources to be exploited, hydroelectricity is the low-hanging fruit of the renewable world. It's steady, self-storing, highly efficient, cost-effective, low-carbon, low-tech, and offers a serious boon to water skiers" (Murphy 2012).

The U.S. has 78 GW of hydroelectric capacity installed. In a year, these plants produce 272 TWh. 0.9 % of the primary energy use, or 2.3 % of primary energy associated with electricity and 7.3 % of the electricity supply is contributed by hydro-electricity.

The energy budget of Earth shows that 23 % of the solar radiation is utilized in evaporating water!

A report by the Eurelectric group assessed the global hydro potential in four cascading steps:

Table 1.1 Some of the studies regarding the importance of location in selection of sites for hydropower plants

Author	Place of work	Algorithms	Limitations
Choong-Sung Yi, Jin-Hee Lee, Myung-Pil Shim	Geum River Basin, Korea	Geo-Spatial Information System (GSIS)	Large area can be precisely surveyed within a short period of time
Pannathat Rojanamon, Taweep Chaisomphob, Thawilwadee Bureekul	Nan river basin in Thailand	Geographic Information System (GIS) technology	Rural and mountainous areas, the large amount of data required, and the lack of participation of the local people living
A.A. Ghadimi, F. Razavi, B. Mohammadian.	Lorestan province in Iran		High transmission costs and insufficient supply
B.C. Kusre, D.C. Baruah, P.K. Bordoloi, S.C. Patra	Kopili River basin in Assam (India)	GIS and hydrological model SWAT2000	Fossil fuels, increasing power demand and availability of untapped water resources

1. Gross potential if all runoff is developed to sea level with no loss;
2. Technical potential, ignoring economic limitations;
3. Economically viable potential, cost competitive with other sources;
4. Exploitable potential, considering environmental and other restrictions.

According to their study results first step yielded a potential of 5.8TW whereas second, third and fourth step estimated a power potential of 4.4, 4.6 and 5.1 TW respectively.

When restrictions of technical feasibility is imposed, these same sources estimate 1.6–2.3 TW globally. In case of the restrictions of Economic feasibility is applied (in today's economic climate) drops this to 1.0–1.4 TW where as Environmental restrictions (in today's climate) reduce this number further.

In this regard the location selection of the sites for hydropower plant becomes highly important as depicted in Table 1.1.

In case of hydropower plants the following factors are found to be highly important in justifying the installations of SHPs (Roberts and Mosey 2013; Tuna 2013; Sharma and Awal 2013):

1. Average Discharge
2. Probability of Flow
3. Annual Variations in the Difference in Water Level
4. Head Duration Curve
5. Potential Power
6. Population Density
7. Hostile Population
8. Utilization Potential

9. Distance from Nearest Grid
10. Distance from Nearest Consumer
11. Slope
12. Area of Forest Cover
13. Area of Agricultural Field
14. Area of Water Body
15. Presence of Wildlife
16. Presence of Endangered Species
17. Tourism Potential
18. Turbulence
19. Amount of Sedimentation
20. Water Quality

All these factors are a function of location and vary with change in the locations even within the same watershed. That is why site selection or feasibility for hydropower plant is always performed in the selected alternatives before the approval of the project. If locations are selected logically the reliability and efficiency of the HPP becomes sustainable.

1.4 Different Decision-Making Algorithms

There are various linear and nonlinear multi-criteria decision making algorithms (Kelly et al. 2013; Mahdi et al. 2002) which are imposed for different field of decision making. Some of the algorithms are described below based on there frequency of application to solve decision support problems.

Neural Network (ANN): The modeling framework which follows the methodology and process flow of human nervous systems is referred as artificial neural network models. The ANN is applied to predict, classify, optimize as well as decide to solve problems from various field of study (Liu et al. 2013; Chang and Sanfey 2013; Wu et al. 1993). The term neural network was traditionally used to refer to a network or circuit of neurons. The modern usage of the term often refers to artificial, which are composed of artificial neurons or nodes.

Genetic Algorithm (*GA*): Genetic algorithms are a part of evolutionary computing, which is a rapidly growing area of artificial intelligence. The genetic algorithm, the crossover, and mutation are the most important part of the genetic algorithm. Although in most of the cases GA is applied for searching the optimal solution but due to the flexibility of the method this algorithm has also been applied in various decision making studies to find for the better alternatives among the many available (Long et al. 2013; Pal et al. 2013; Fernandez et al. 2013).

Fuzzzy decision making: In all aspects of multi-criteria decision making (MCDM) the theory and practice of fuzzy optimization and decision making is applied widely (Kannan et al. 2013; Villarroel et al. 2013). The logic utilizes the theory of fuzziness to examine theoretical, empirical, and experimental work related to various decision support system employed to solve related problems.

Analytical Hierarchy Process (*AHP*): The analytic hierarchy process is a structured technique for organizing and analyzing complex decisions. It has particular application in group decision making and is used around the world in a wide variety of decision situations in fields such as government, business, industry, healthcare, and education (Rao 2013; Nor et al. 2013; Srdjevic et al. 2013).

1.5 Objective and Brief Methodology

The main objective for the present investigation is to analyze the potential for common decision-making algorithms as described in the earlier section in selection of suitable locations for Large-scale hydropower plants (LHP).

The efficiency of SHP depends largely on the location of the plant as geophysical properties vary with location and such properties influence the potential of hydropower in that region.

Also, before installation of hydropower plants the environmental and socioeconomic feasibility is required to be verified as both the factors are extremely important for the success of such projects. Again, both these factors are location dependent.

In some places due to the environmental characteristics of the place, the local population will be hostile to hydropower projects due to the requirements of relocation and ecological and environmental destructions (both of which directly impact the socioeconomic status of the population). In some other places, due to the unemployability and scarcity of natural resources, the local mass may demand employment and scope of sustainable livelihood.

That is why the selection of location is important for successful implementations of SHPs.

The present investigation has applied; due to the potential and efficiency of different nature-based algorithms in "learning" complex, nonlinear interrelationships between diverse kinds of variables; common nature-based meta-heuristics to select suitable location for proposed LHPs.

In this aspect three locations were selected as alternatives which are situated in three different countries with different potential of hydropower generation. The River Chenab in India which has already installed hydropower plant generating electricity regularly; River Danube in Germany which has potential of hydropower but not as much as River Chenab, and River Yukon in Alaska where no scope of such kind of power plants exist.

Performance metrics like Coefficient of Agreement was utilized to verify the potential of nature-based algorithms in predicting the suitable alternative for installations of LHPs based on the geophysical, environmental, ecological, and socioeconomical conditions of the three alternatives.

1.6 Locations Considered as Alternatives

The locations for the present investigation are selected in such a way that with the help of the same, the performance of the decision-making algorithms can be compared. In the study upstream of River Chenab in India, upstream of River Danube in Germany, and downstream of River Yukon in Alaska are selected as the study areas.

1.6.1 The Chenab

The source of Rive Chenab is from Bara Lacha Pass, 32°44′N 77°26′E, in Himachal Pradesh. The waters flowing south from the pass are known as the Chandra River and those that flow north are called the Bhaga River. Eventually, the Bhaga and Chandra river merges at the village of Tandi. It becomes the Chenab when it joins the Marau River at Bhandera Kot, 12 km from Kishtwar Town in Jammu and Kashmir (Fig. 1.2).

It flows from the Jammu region of Jammu and Kashmir into the plains of the Punjab, forming the boundary between the Rechna and Jech interfluves (Doabs in Persian). It is joined by the Jhelum River at Trimmu تریمو and then by the Ravi

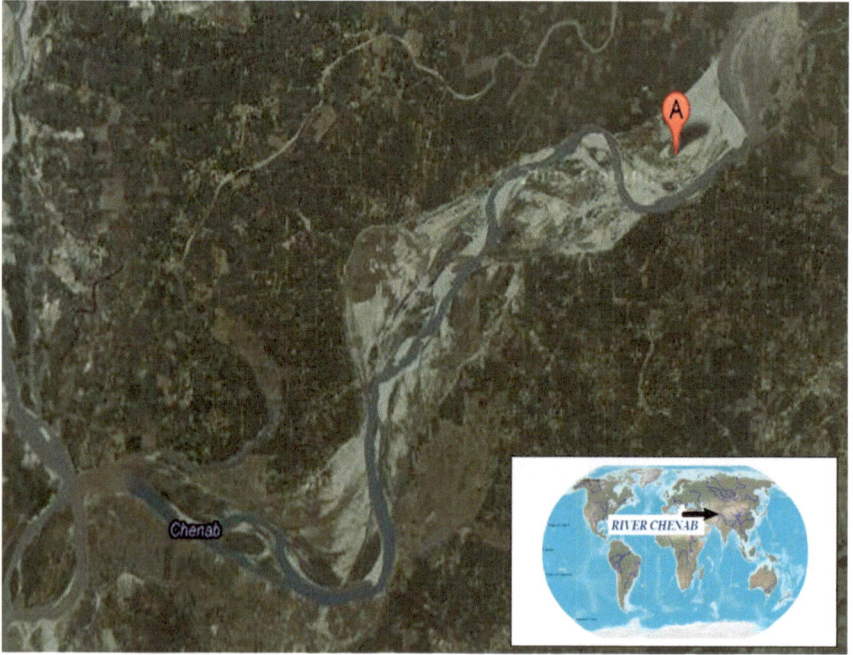

Fig. 1.2 The course of River Chenab in India (*Courtesy* Google Earth)

River. It then merges with the Sutlej River near Uch Sharif, Pakistan to form the Panjnad or the 'Five Rivers,' the fifth being the Beas River which joins the Satluj near Ferozepur, India. The Chenab then joins the Indus at Mithankot. The total length of the Chenab is approximately 960 km.

The flow path of River Chenab is through the alluvial plains of the Punjab province in Pakistan for a distance of 3,398 miles.

Total Catchment Area: 26,035 miles^2
Annual Average Flow: 12.38 MAF (10.07 Kharif and 2.31 Rabi)

1.6.2 The Danube

The Danube is a river in Central Europe which is Europe's second longest after the Volga. The River is classified as an international waterway, the source of the river is Donaueschingen which is located at Black Forest. The river passes through Romania (29.0 % of basin area), Hungary (11.6 %), Serbia (10.2 %), Austria (10.0 %), Germany (7.0 %), Slovakia (5.9 %), Bulgaria (5.9 %), Croatia (4.4 %), Ukraine (3.8 %), and Moldova (1.6 %)

Length: 2,860 km
Discharge: 6,500 m/s
Source: Black Forest
Location: Germany
Average flow: 6,500 m^3/s (229,545 ft^3/s)

The watershed of River Danube has an area of 315,000 miles^2 (817,000 km^2) includes a variety of natural conditions that affect the origins and regimes of its watercourses. The river includes some 300 tributaries, more than 30 of which are navigable. The river basin expands unevenly along its length. It covers about 18,000 miles^2 (47,000 km^2) at the Inn confluence, 81,000 miles^2 (210,000 km^2) after joining with the Drava, and 228,000 miles^2 (590,000 km^2) below the confluences of its most affluent tributaries, the Sava and the Tisza. The right-bank tributaries, is the major contributors (nearly two-thirds of total river runoff) which collects water from the Alps to main course of River Danube which collect their waters from the Alps and other mountain areas and contribute up to two-thirds of the total river runoff or outfall.

The river can be divided into three stretches. The upper course, called the Hungarian Gates, in the Austrian Alps and the Western Carpathian Mountains includes the course of river from the source to the gorge. The middle course runs from the Hungarian Gates to the Iron Gate Gorge in the Southern Romanian Carpathians. The lower course flows from the Iron Gate to the delta-like estuary at the Black Sea (Fig. 1.3).

Fig. 1.3 The course of River Danube (*Courtesy* Google Maps™)

The upper Danube springs have two small tributaries—the Breg and Brigach—from the eastern slopes of the Black Forest mountains of Germany, which partially consist of limestone and from Donaueschingen, where the headstreams unite. After the confluence, the river takes the northeastward direction through narrow, rocky bed. In the north rises the wooded slopes of the Swabian and Franconian mountains; in between of which the river forms a scenic canyon-like valley. The south of the river course stretches the large Bavarian Plateau, covered with thick layers of river deposits from the numerous Alpine tributaries. The bank is low and uniform, composed mainly of fields, peat, and marshland.

At Regensburg the Danube reaches its northernmost point, from which it veers south and crosses wide, fertile, and level country. Shortly before it reaches Passau on the Austrian border, the river narrows and its bottom abounds with reefs and shoals. The Danube then flows through Austrian territory, where it cuts into the slopes of the Bohemian Forest and forms a narrow valley. In order to improve navigation, dams and protecting dikes have been built near Passau, Linz, and Ardagger. In the upper Danube, some 600 miles (965 km) long, has a considerable average inclination of the riverbed (0.93 %) and a rapid current of 2–5 miles/h. Depths vary from 3 to 26 ft (1–8 m). The Danube expands substantially at Passau where the Inn River, its largest upstream tributary, merges carrying more water than the main river. Other major tributaries in the upper Danube course include the Iller, Lech, Isar, Traun, Enns, and Morava rivers.

The middle course of the Danube looks more like a flatland river, with low banks and a bed that reaches a width of more than one mile. Only in two sectors—at Visegrád (Hungary) and the Iron Gate—does the river flow through narrow, canyon-like gorges. The basin of the middle Danube depicts two main features—the flatland of the Little Alfold and the Great Alfold plains, and the low peaks of the Western Carpathians and the Transdanubian Mountains.

The Danube enters the Little Alfold plain immediately after the Hungarian Gates Gorge near Bratislava, Slovakia. There the river stream slows down abruptly and loses its transporting capacity, so that enormous quantities of gravel and sand settle on the bottom. East of Komárno the Danube enters the Visegrád Gorge, in between the foothills of the Western Carpathian and the Hungarian Transdanubian Mountains.

1.6.3 The Youkon

The Yukon River (Fig. 1.4) is a major river of northwestern North America. The source of the river is located in British Columbia, Canada. The next stretch lies in, and gives its name to the Yukon territory. Its overall length is 3,187 km, with 1,149 km within Canadian borders. The watershed's total drainage area is 840,000 km^2 (323,800 km^2 in Canada).

Length: 3,187 km
Discharge: 6,428 m/s
Source: Atlin Lake
Location: North America
Average flow: 6,430 m/s (227,000 ft/s).

Fig. 1.4 The course of River Yukon in Alaska (*Courtesy* Google Map™)

The lower half of the river lies in the U.S. state of Alaska. The river is 3,190 km long and empties into the Bering Sea at the Yukon-Kuskokwim Delta. The total drainage area is 832,700 km, of which 323,800 km is in Canada. Which is more than 25 % larger than Texas or Alberta.

Paddle-wheel riverboats continued to navigate through the river until the 1950s, when the Klondike Highway was completed.

Yukon means "great river" in Gwich'in. The river was also called Kuigpak, or "big river," in Central Yup'ik. The Lewes River is the former name of the upper course of the Yukon, from Marsh Lake to the confluence of the Pelly River at Fort Selkirk.

Although in the watershed of Yukon river there are many gold mines, military installations, dumps, wastewater and other sources of pollution the water quality data of the river shows relatively good levels of turbidity, metals and dissolved oxygen according to the US Geological Survey data.

The Yukon River Inter-Tribal Watershed Council, a cooperative effort of 64 First Nations and tribes in Alaska and Canada, has the goal of making the river and its tributaries safe to drink from again, by supplementing and scrutinizing Gov-389 ernment data.

In the present investigation the downstream to White Course was considered as an alternative for location selection for a large-scale hydropower plant.

References

W. Beckerman, Economic growth and the environment: whose growth? Whose environment? World Dev. **20**(4), 481–496 (1992)

R.S. Bradley, M. Vuille, H.F. Diaz, W. Vergara, Threats to water supplies in the tropical Andes. Science **312**(5781), 1755–1756 (2006)

L.J. Chang, A.G. Sanfey, Great expectations: neural computations underlying the use of social norms in decision-making. Soc. Cogn. Affect. Neurosci. **8**(3), 277–284 (2013)

T. Davies, Introduction, An. EPA J. **16**, 2 (1990)

E. Fernandez, E. Lopez, G. Mazcorro, R. Olmedo, C.A. Coello Coello, Application of the non-outranked sorting genetic algorithm to public project portfolio selection. Int. J. Inf. Sci. **228**, 131–149 (2013)

K.D. Frederick, D.C. Major, Climate change and water resources. Clim. Change **37**(1), 7–23 (1997)

J. Goldemberg, Ethanol for a sustainable energy future. Science **315**(5813), 808–810 (2007)

P. Hanlon, R. Madel, K. Olson-Sawyer, K. Rabin, J. Rose, K. Demaline, K. Sweetman, Food Water Energy: Know the Nexus, GRACE Communications Foundations, Retrieved from http://www.gracelinks.org/media/pdf/knowthenexus_final.pdf (2013)

International Energy Agency, World Energy Outlook 2012, (ISBN 978-92-64-18084-0), (2012)

M. Jaccard, Sustainable Fossil Fuels: The Unusual Suspect in the Quest for Clean and Enduring Energy (Cambridge University Press, Cambridge, 2006)

C.M. Kelly, E.F. Hequet, J.K. Dever, Breeding for improved yarn quality: modifying fiber length distribution. Ind. Crops Prod. **42**, 386–396 (2013)

W.F. Laurance, M.A. Cochrane, S. Bergen, P.M. Fearnside, P. Delamônica, C. Barber, S. D'Angelo, T. Fernandes, Environment: the future of the Brazilian Amazon. Science **291** (5503), 438–439 (2001)

W. Liu, L. Quan, A multi-criteria decision making method based on linguistic preference information for IT outsourcing vendor selection in hospitals, Advances in Intelligent Systems Research (2013). doi:10.2991/icibet.2013.65

J. Liu, Y. Wu, T. Tao, Q. Chu, Research and development of decision support system for regional agricultural development programming. In computer and computing technologies in agriculture VI. (Springer Berlin Heidelberg, 2013), pp. 271–281

L. Lohmann, *Carbon Trading. A Critical Conversation on Climate Change, Privatisation and Power* (Dag Hammarskjöld Foundation, Uppsala, 2006)

C. Long, J. Li, H. Dong. A modeling of the description of urban residents traveling decision based on simple genetic algorithm. In LISS 2012. (Springer Berlin Heidelberg, 2013), pp. 679–683

D. Kannan, R. Khodaverdi, L. Olfat, A. Jafarian, A. Diabat, Integrated fuzzy multi criteria decision making method and multi-objective programming approach for supplier selection and order allocation in a green supply chain. J. Clean. Product. (2013)

I.M. Mahdi, M.J. Riley, S.M. Fereig, A.P. Alex, A multi-criteria approach to contractor selection. Eng. Constr. Archit. Manag. 9(1), 29–37 (2002)

S. Mukherjee, C. Debashis, Negative Influence of Fiscal Subsidies on Environment: Empirical Evidence from Cross-Country Estimation. No. 13/117 (2013)

T. Murphy, Potential of hydro-electricity a detailed look at hydroelectricity's potential, http://oilprice.com/Alternative-Energy/Hydroelectric/A-Detailed-Look-At-Hydroelectricitys-Potential.html. Accessed 2012

M. Nor, N. Hisyamudin, C.K. Sia, L.Y. Lee, S.H. Chong, M.A. Azlan. Decision making with the analytical hierarchy process (AHP) for material selection in screw manufacturing for minimizing environmental impacts. App. Mech. Mat. 315, 57–62 (2013)

B.B. Pal, A. Mukhopadhyay, P. Biswas, S. Mukherjee, Using fuzzy goal programming to solve electric power generation and dispatch problem via genetic algorithm. (2013)

D. Pimentel, J. Houser, E. Preiss, O. White, H. Fang, L. Mesnick, S. Alpert, Water resources: agriculture, the environment, and society. BioScience 47, 97–106 (1997)

R.V. Rao, Applications of improved MADM methods to the decision making problems of manufacturing environment. In decision making in manufacturing environment using graph theory and fuzzy multiple attribute decision making methods. (Springer, London, 2013), pp. 41–135

J.O. Roberts, G. Mosey, Feasibility study of economics and performance of a hydroelectric installation at the jeddo mine drainage tunnel. A study prepared in partnership with the environmental protection agency for the RE-powering america's land initiative: siting renewable energy on potentially contaminated land and mine sites. No. NREL/TP-6A20-51321. National Renewable Energy Laboratory (NREL)'Golden, CO., (2013)

R.H. Sharma, R. Awal. Hydropower development in Nepal. Renew. Sust. Energ. Rev. 21, 684–693 (2013)

B.K. Sovacool, Contextualizing avian mortality: a preliminary appraisal of bird and bat fatalities from wind, fossil-fuel, and nuclear electricity. Energ. Policy 37(6), 2241–2248 (2009)

Z. Srdjevic, M. Lakicevic, B. Srdjevic, Approach of decision making based on the analytic hierarchy process for urban landscape management. Environ. Manag. 1–9 (2013)

Triple A Learning, The threat to sustainability from the use of fossil fuels, Microeconomics, http://www.chatt.hdsb.ca/~dalyr/FOV1-000DC265/FOV1-0013E47C/FOV1-0013E47D/page_181.htm

M.C. Tuna, Feasibility assessment of hydroelectric power plant in ungauged river basin: A case study. Arabian J. Sci. Eng. 1–9 (2013)

F. Villarroel, J. Espinoza, C. Rojas, J. Rodriguez, M. Rivera, D. Sbárbaro, Multiobjective switching state selector for finite states model predictive control based on fuzzy decision making in a matrix converter. 1–1 (2013)

P.M. Vitousek, H.A. Mooney, J. Lubchenco, J.M. Melillo, Human domination of Earth's ecosystems. Science **277**(5325), 494–499 (1997)

Y. Wu, M.L. Giger, K. Doi, C.J. Vyborny, B.A. Schmidt, C.E. Metz, Artificial neural networks in mammography: application to decision making in the diagnosis of breast cancer. Radiology **187**(1), 81–87 (1993)

Chapter 2
Hydropower Plants

Abstract The hydropower plants can be broadly subdivided into different classes based on quantity of water available, available head, and nature of load. Based on quantity of water available hydro power plant can be classified into Reservoir plants, Run-off river plants with pondage, and Run-off River plants without pondage. In case of available head, hydropower plants can be further subdivided into Low head, Medium head, and High head. With respect to nature of Load any hydropower plant can be grouped into Base and Peak load plants.

2.1 Classification of Hydropower Plants

The classification of hydro-electric plants is based upon (Fig. 2.1):

(a) Quantity of water available
(b) Available head
(c) Nature of load

2.1.1 Classification with Respect to Quantity of Water Available

(i) Run-off river plants without poundage: These plants does not have storage or pondages to store water; Run-off river plants without pondages uses water as it comes. The plant can use water as and when available. Since, generation capacity of these type of plants these plants depend on the rate of flow of water, during rainy season high flow rate may mean some quantity of water to go as waste while during low run-off periods, due to low flow rates, the generating capacity will be low.

M. Majumder and S. Ghosh, *Decision Making Algorithms for Hydro-Power Plant Location*, SpringerBriefs in Energy, DOI: 10.1007/978-981-4451-63-5_2,
© The Author(s) 2013

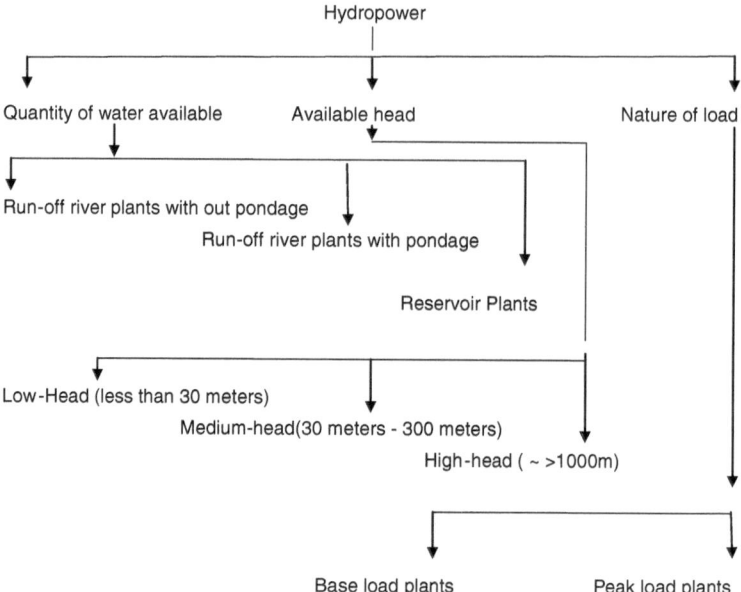

Fig. 2.1   Showing the classification overview of hydro-power plants

(ii) Run-off river plants with pondage: In these plants, pondage allows storage of water during lean periods and use of this water during peak periods. Based on the size of the storage structure provided it may be possible to cope with hour-to-hour fluctuations. This type of plant can be used on parts of the load curve as required, and is more useful than a plant without pondage. If pondage is provided, tail race conditions should be such that floods do not raise tail-race water level, thus reducing the head on the plant and impairing its effectiveness. This type of plant is comparatively more conscientious and its generating capacity is unabased on available rate of flow of water.

(iii) Reservoir plants: A reservoir plant is that which has a reservoir of such size as to accede carrying over storage from wet season to the next dry season. Water is stored behind the dam and is available to the plant with control as required. This type of plant has better extent and can be used efficiently throughout the year. Its firm capacity can be expanded and can be utilized either as a base load plant or as a peak load plant as required. It can also be used on any portion of the load curve as required. Maximum hydro-electric plants are of this type.

2.1.2 Classification According to Availability of Water Head

(i) Low-head (less than 30 m) hydro-electric plants: "Low head" hydro-electric plants are power plants which generally utilize heads of only a few meters or less. Power plants of this type may utilize a low dam or weir to channel water, or no dam and simply use the "run of the river". Run of the river generating stations cannot store water, thus their electric output varies with seasonal flows of water in a river. A large volume of water must pass through a low head hydro plant's turbines in order to produce a useful amount of power. Hydro-electric facilities with a capacity of less than about 25 MW (1 MW = 1,000,000 W) are generally referred to as "small hydro", although hydro-electric technology is basically the same regardless of generating capacity.

(ii) Medium-head (30–300 m) hydro-electric plants: These plants consist of a large dam in a mountainous area which creates a huge reservoir. The Grand Coulee Dam on the Columbia River in Washington (108 m high, 1,270 m wide, and 9,450 MW) and the Hoover Dam on the Colorado River in Arizona/Nevada (220 m high, 380 m wide, and 2000 MW) are good examples. These dams are true engineering marvels. In fact, the American Society of Civil Engineers as designated Hoover Dam as one of the seven civil engineering wonders of the modern world, but the massive lakes created by these dams are a graphic example of our ability to manipulate the environment—for better or worse. Dams are also used for flood control, irrigation, recreation, and often are the main source of potable water for many communities. Hydro-electric development is also possible in areas such as Niagra Falls where natural elevation changes can be used.

(iii) High-head hydro-electric plants: "High head" power plants are the most common and generally utilize a dam to store water at an increased elevation. The use of a dam to impound water also provides the capability of storing water during rainy periods and releasing it during dry periods. This results in the consistent and reliable production of electricity, able to meet demand. Heads for this type of power plant may be greater than 1,000 m. Most large hydro-electric facilities are of the high-head variety. High-head plants with storage are very valuable to electric utilities, because they can be quickly adjusted to meet the electrical demand on a distribution system.

2.1.3 Classification According to Nature of Load

(i) Peak Load Plants : The peak load plants are used to supply power at the peak demand phase. The pumped storage plants and Gas Turbine plants are this type of plants. Their efficiency varies between 60–70%.

(ii) Base load plants: A base load power plant is one that provides a steady flow of power regardless of total power demand by the grid. These plants run at all times through the year except in the case of repairs or scheduled maintenance.

2.2 Advantages of Hydroelectric Plants

The benefits of hydropower plants are manifold as described below:

- The running, operation and maintenance cost of this kind of plants are low.
- After the initial infrastructures are developed the energy is virtually free.
- The plants is totally free of pollution as no conventional fuels are required to be burned.
- The lifetime of generating plants are substantially long.
- Reliability is much more than wind, solar or wave power due to its easy availability and convertibility.
- Water can be stored above the dam ready to cope with peaks in demand.
- The uncertainties that arises due to unscheduled breakdowns are relatively infrequent and short in duration due to the simplicity and flexibility of the instruments.
- Hydro-electric turbine generators can be started and put "on-line" very rapidly.
- It is possible to produce electricity from hydro-electric power plant if flow is continuously available. Benkovic et al. (2013); Panic et al. (2013); Sharma and Awal (2013); Dursun and Cihan (2011) etc. has already discussed the benefits of the HPP in different aspect in their published literatures.

2.3 Disadvantages of Hydro-Electric Plants

But along with the advantages, the disadvantages (Bahadori et al. 2013; Chen et al. 2013; Jensen et al. 2002) of such renewable energy projects are also manifold but lesser than the other sources of infinite and also finite energy.

- The potential of hydro power depend on locations and if properly not selected may cause lots of hostility and absurdity during operational stage of the power plant.
- The dams are very expensive to build. However, many dams are also used for flood control or irrigation, so building costs can be shared.
- The capital cost of electrical instruments along with civil engineering works to be installed and cost of laying transmission lines is generally high.
- The impact on plant life due to the water quality and quantity downstream of hydro power plants are reported.
- The impact on residents and the environment may be unacceptable environmental and social activist if location is not optimally selected.
- Due to increase in water temperature and insertion of excess nitrogen into water at spillways health and migration of fish as well as other aquatic plants get effected.
- Due to the installation of reservoir in the flow paths the siltation rate get altered.

References

A. Bahadori, Z. Gholamrez, Z. Sohrab, An overview of Australia's hydropower energy: Status and future prospects. Renew. Sust. Energ. Rev. **20**, 565–569 (2013)

S. Benkovic, M. Nikola, J. Sandra, Possibilities for development of the Electric Power Industry of Serbia through private source financing of small hydropower plants. Renew. Energ. **50**, 1053–1059 (2013)

X. Chen, W. Zhenyu, H. Sanfeng, L. Fuqiang, Programme management of world bank financed small hydropower development in Zhejiang Province in China. Renew. Sust. Energ. Rev. **24**, 21–31 (2013)

B. Dursun, G. Cihan, The role of hydroelectric power and contribution of small hydropower plants for sustainable development in Turkey. Renew. Energ. **36**(4), 1227–1235 (2011)

S.G. Jensen, S. Klaus, Interactions between the power and green certificate markets. Energ. Policy **30**(5), 425–435 (2002)

M. Panić, U. Marko, M.P. Ana, B. Jovana, B. Željko, Small hydropower plants in Serbia: Hydropower potential, current state and perspectives. Renew. Sust. Energ. Rev. **23**, 341–349 (2013)

R.H. Sharma, A. Ripendra, Hydropower development in Nepal. Renew. Sust. Energ. Rev. **21**, 684–693 (2013)

Chapter 3
Decision Making Methodology

Abstract In case of multiple alternatives or solutions to a given problem a decision maker often gets confused to select the ideal option which will yield the maximum benefit. That is why there are various algorithms which assist in logical and scientific decision making so that the costs can be reduced and in the same time benefits can be multiplied without compromising the basic requirement. The procedure of scientific and logical decision making involves fixing a goal, defining and selecting the criteria and alternatives, and applying decision making tools to make a decision after comparing all the alternatives with respect to each of the alternatives. Multicriteria Decision Making is nowadays gaining popularity in many field of research as solution of most of the problems came from taking an optimal decision. A suitable but logical and practical procedure to identify alternatives can reduce many discrepancy and inefficiency of a probable solution.

The rational decision making becomes necessary when there are either various alternatives (Multi-Criteria Decision Analysis) or a need to identify the optimal characteristics of related criteria so that an optimal decision can be adopted for maximum benefit.

Decision making is the science of identification and selection of alternatives based on the values and preferences of a decision maker. A need of taking a decision arises whenever there are multiple options or solutions to a given problem and in such a case the objective is not only to identify as many of these alternatives as possible but "to choose the one that best fits with our goals, objectives, desires, values, and so on" (Harris 1980).

3.1 Procedures of Decision Making

According to Baker et al. (2001), decision making starts with the identification of the decision maker(s) and stakeholder(s) in the decision which reduces the possible disagreement about problem definition, requirements, goals and criteria. A general decision making process can be divided into the following steps:

M. Majumder and S. Ghosh, *Decision Making Algorithms for Hydro-Power Plant Location*, SpringerBriefs in Energy, DOI: 10.1007/978-981-4451-63-5_3,

Step 1. Define the problem

This action must, as a minimum, recognize root causes, satisfying all the assumptions, system, organizational boundaries and interfaces, along with any stakeholder issues.

The goal is to convey the issue in a clear, one-sentence problem statement that describes both the initial conditions and the desired conditions.

Of course, in case of complex decision making problems the one-sentence limit is often exceeded in the practice.

The problem statement has to be concise and unambiguous written material agreed by all decision makers and stakeholders.

Step 2. Determine requirements

Requirements are conditions that any acceptable solution to the problem must meet.

In mathematical form, these exigencies are the constraints specifying the set of feasible solutions of the decision problem.

Step 3. Establish goals

The goal of a decision making represents the minimum essential requirements that the decision must satisfy.

In mathematical form, the goals are objectives contrary to the compulsions that are constraints.

In practical decision making it is common to have conflicting goals.

Step 4. Identify alternatives

A problem with various preferences offer different approaches for supplanting the initial condition into the desired condition.

The number of possible alternatives may be finite or infinite for a given problem. In case of the former the preferences can be evaluated as per the goal and requirements of the decision making problem and the infeasible ones may be deleted out from further considerations. Ultimately after the operation an explicit list of alternatives can be obtained.

The problems with infinite solutions can be treated as mathematical optimization problems where the solutions which satisfy the mathematical form of constraints becomes the most selectable solutions which has the potentiality to achieve the desired goals.

Step 5. Define criteria

Criteria are the factors/parameters/deciders which implode a change in the identity/occurrence of the objective variable. A change in the criteria will influence the selection of alternatives as the most optimal solution for a given problem among all the alternatives considered. According to Baker et al. (2001), criteria can

- discriminate among the preferences to support the comparison of the performance of the alternatives,
- complete to include all goals,
- functional.

  The complexity of decision making can be reduced by the removal of the redundant, highly correlated and duplicate criteria with the help of different criteria selection methods (Delphi Kannan et al. 2013; Li et al. 2002; Mahdi et al. 2002; Least Square Method Park et al. 2013), Min–Max Deviation (Cheng et al. 2013 etc.). The system of criteria must represent the following characteristics

(a) Systemic principles: The criterias must represent essential characteristic and performance of the system.
(b) Consistent: They must be consistent with the decision making objectives.
(c) Independent: The criterias must not have any relation with other same level criterias.
(d) Mensurability: Criterias should be measurable either by quantity or quality.
(e) Comparability: The dimensions of the criterias must be removed to compare them in a common reference frame.

Step 6. Select a decision making tool
The selection of an appropriate tool depends on the concrete decision problem, as well as on the objectives of the decision makers. Sometimes the "simpler the method, the better" but complex decision problems may require complex methods, as well.

Step 7. Evaluate alternatives against criteria
An ideal method for decision making requires, as input data, the appraisals of the alternatives against the criteria. Depending on the criterion, the assessment may be objective, with respect to some commonly shared and understood scale of measurement or can be subjective, representing the subjective assessment of the evaluator.

After the assessment the selected decision making tool can be applied to rank the recourses or to choose a subset of the most promising druthers.

Step 8. Validate solutions against problem statement
The alternatives selected by the applied decision making tools is required to be validated against the requirements and goals of the decision problem. It may happen that the decision making tool was misapplied. In complex problems the selected prerogatives may also invite the attention of the decision makers and stakeholders that further goals or requirements should be added to the decision model.

3.2 Multi-Attribute Decision Making

In a given problem when the number of the criteria and preferences are finite, and the alternatives are given explicity then this type of problems are called as multi-attribute decision making problems.

Consider a multi-attribute decision making problem with p criteria and q alternatives. Let $C_1, \ldots, C_{pm}$ and $A_1, \ldots, A_{qn}$ denote the criteria and alternatives, respectively.

A standard feature of multi-attribute decision making methodology is the decision table.

Table 3.1 Table showing some of the studies related to decision making algorithms

Authors name	Algorithms used	Objective	Problems/ Limitations
Yong-Sheng Ding, Zhi-Hua Hu, Wen-Bin Zhang	Multi-criteria decision making (MCDM) model	Size fitting problem is a main obstacle to large scale garment sales and online sales because it is difficult to find the fit garments by the general size information	Garment matching problem
M.J. Chen, G.H. Huang	Derivative algorithm (DAM)	Environmental systems analysis	The Inexact quadratic programming (IQP) application to large-scale problems
S. Martorell, J.F. Villanueva, S. Carlos, Y. Nebot, A. Sánchez, J.L. Pitarch, V. Serradell	Genetic algorithm	Technical specifications and maintenance (TSM) activities at nuclear power plants (NPP)	Improved level of plant safety
L. Castillo, C.A. Dorao	Binary genetic algorithm	The oil and gas projects based on game theory	Soft information with large uncertainties

In the table each row belongs to a criterion and each column depicts the performance of an alternative.

The score a_{ij} describes the performance of alternative A_j against criterion C_i.

Weight w_i reflects the relative importance of criteria C_i to the decision, and is assumed to be positive.

The weights of the criteria are usually determined on subjective basis.

They represent the opinion of a single decision maker or synthesize the opinions of a group of experts using a group decision technique, as well. Table 3.1 shows some of the applications of MCDA in solving real life problems.

3.3 Performance Metrics: KAPPA Coefficient of Agreements

Kappa assesses the percentage of data values in the main diagonal of the agreement matrix and then these values are adjusted for the amount of agreement that could be expected due to chance alone.

Cohen's kappa measures the agreement between two raters who each classify N items into C mutually exclusive categories. Galton (1892) used KAPPA for the first time (Smeeton 1985).

The equation for κ is:

$$\kappa = \frac{\Pr(a) - \Pr(e)}{1 - \Pr(e)}, \tag{3.1}$$

where $\Pr(a)$ is the relative observed agreement among raters, and $\Pr(e)$ is the hypothetical probability of chance agreement, using the observed data to determine the probabilities of each observer randomly saying each category. If the classifiers are in complete agreement then $\kappa = 1$. If there is total disagreement among the raters other than what would be expected by chance (as defined by $\Pr(e)$), $\kappa = 0$.

The maximum value for kappa occurs when the observed level of agreement is 1, which makes the numerator as large as the denominator. As the observed probability of agreement declines, the numerator declines.

3.3.1 Interpretation of KAPPA Coefficient of Agreement

Kappa is always less than or equal to 1. In rare situations, Kappa can be negative. This is a sign that the two observers agreed less than would be expected just by chance.

It is rare that we get perfect agreement. Different people have different interpretations as to what is a good level of agreement. Here is one possible interpretation of Kappa (Altman 1991).

Poor agreement = Less than 0.20
Fair agreement = 0.20 to 0.40
Moderate agreement = 0.40 to 0.60
Good agreement = 0.60 to 0.80
Very good agreement = 0.80 to 1.00

KAPPA has been utilized in various studies to evaluate the performance of analytical and statistical classification and clusterization methods (Sexton et al. 2013; Vanonckelen 2013; Cheng and Deanna 2013).The KAPPA metrics is satisfactorily utilized in Multi Criteria Decision Analysis to verify the accuracy of decision making methods by comparing an already decided situation with the output of the decision making algorithm (Castillo 2013; Ghadimi et al. 2011; Kusre et al. 2010; Martorell et al. 2005; Rojanamon et al. 2009; Yang 2010; Yi et al. 2010).

References

D.G.Altman, *Practical statistics for medical research* (Chapman and Hal, London,1991)

L. Castillo, C.A. Dorao, Decision-making in the oil and gas projects based on game theory: Conceptual process design. Energy Convers. Manage. **66**, 48–55 (2013)

H. Cheng, H. Weilai, Z. Quan, C. Jianhu, Solving fuzzy multi-objective linear programming problems using deviation degree measures and weighted max–min method. Appl. Math. Model. (2013)

Y. Cheng, L.M. Deanna, Classification accuracy and consistency of computerized adaptive testing. Behav. Res. Methods. **45**(1), 132–142 (2013)

F. Galton, *Finger prints*. (Macmillan and Company, 1892)

A.A. Ghadimi, F. Razavi, B. Mohammadian, Determining optimum location and capacity for micro hydropower plants in Lorestan province in Iran. Renew. Sustain. Energy Rev. **15**(8), 4125–4131 (2011)

D. Kannan, K. Roohollah, O. Laya, J. Ahmad, D. Ali, Integrated fuzzy multi criteria decision making method and multi-objective programming approach for supplier selection and order allocation in a green supply chain. J. Clean. Prod. (2013)

B.C. Kusre, D.C. Baruah, P.K. Bordoloi, S.C. Patra, Assessment of hydropower potential using GIS and hydrological modeling technique in Kopili River basin in Assam (India). Appl. Energy **87**(1), 298–309 (2010)

Li, Shuliang, Barry Davies, John Edwards, Russell Kinman, and Yanqing Duan. Integrating group Delphi, fuzzy logic and expert systems for marketing strategy development: the hybridisation and its effectiveness.Market. Intell. Plann. **20**(5), 273–284 (2002)

M.I. Mahdi, J.R. Mike, M.F. Sami, P. Alex, A multi-criteria approach to contractor selection.Eng.constr.archit.manag **9**(1), 29–37 (2002)

S. Martorell, J.F. Villanueva, S. Carlos, Y. Nebot, A. Sánchez, J.L. Pitarch, V. Serradell, RAMS+C informed decision-making with application to multi-objective optimization of technical specifications and maintenance using genetic algorithms. Reliab. Eng. & Syst. Saf. **87**(1), 65–75 (2005)

H. Park, S. Fumitake, K. Sadanori, Robust sparse regression and tuning parameter selection via the efficient bootstrap information criteria. J. Stat. Comput. Sim. 1–12 (2013) [epub ahead of print]

P. Rojanamon, T. Chaisomphob, T. Bureekul, Application of geographical information system to site selection of small run-of-river hydropower project by considering engineering/economic/ environmental criteria and social impact. Renew. Sustain. Energy Rev. **13**(9), 2336–2348 (2009)

O.J. Sexton, D.L. Urban, M.J. Donohue, S. Conghe, Long-term land cover dynamics by multi-temporal classification across the Landsat-5 record. Remote Sens. Environ. **128**, 246–258 (2013)

N.C. Smeeton, Early History of the Kappa Statistic. Biometrics **41**, 795 (1985)

S. Vanonckelen, L. Stefaan, V.R. Anton, The effect of atmospheric and topographic correction methods on land cover classification accuracy. Int. J. Appl. Earth. Obs. Geoinf. **24**, 9–21 (2013)

X.-S. Yang, in *A New Metaheuristic Bat-Inspired Algorithm, in Nature Inspired Coop-erative Strategies for Optimization (NISCO 2010)* , ed. by J.R. Gonzalez et al. Studies in Computational Intelligence, vol. 284 (Springer, Berlin, 2010), pp. 65–74

C.-S. Yi, J.-H. Lee, M.-P. Shim, Site location analysis for small hydropower using geo-spatial information system. Renewable Energy **35**(4), 852–861 (2010)

Chapter 4
Nature-Based Algorithms

Abstract Nature-based algorithms are those algorithms which mimics nature to solve a real-life problem. This kind of meta-heuristic algorithms are popular to search for an optimal answer within a given set of nonlinear complex problems by replicating the way by which nature is solving its problems. For example, bat search for food with the help of the emitted sonar signal which accurately identifies the location of ideal sources of food. This same concept can be replicated in case of real-life problems to estimate the ideal solution of nonlinear problems. In this chapter, the concepts of neural network, fuzzy logic, bat algorithm, and Analytical Hierarchy Process which are applied in the present study for taking a scientific decision in regard to find a suitable location for hydropower plants.

4.1 Artificial Neural Networks

Nature is an active solver of nonlinear and complex problems in result of which the natural equilibrium still exists along with the living beings. For any problem like identification of food locations for animals like BAT nature has devised the sonar signals and its variations which helps the animal to identify a potential source of food. Similarly human nervous system is a signal flow network through which signals are continuously flowing to provide answers to the questions arised by the sensory organs or receptor. The response to the stimuli creates a signal which is processed according to experience and a result is yielded to mitigate the query. Such solution methodology are also adapted in the problems of complex interactions of practical world. Various models to solve nonlinear and complex problems already mimics the way of the nature to solve somewhat similar kind of problems. The major nature based algorithms which are popular than the other methods are utilized are discussed below.

Artificial neural networks have changed the way researchers solve various complex and real-world problems in engineering, science, economics, and finance.

M. Majumder and S. Ghosh, *Decision Making Algorithms for Hydro-Power Plant Location*, SpringerBriefs in Energy, DOI: 10.1007/978-981-4451-63-5_4, © The Author(s) 2013

In mimicking a synthetic functional model from the biological nerve cell has three basic components. The synapses of biological nerve cell are encoded as weights. The junction of the biological nerve cell the node in the mimicked network is the one which interconnects the neural network and gives the strength of the association.

A negative weight depicts a repressive affiliation, whereas positive values designate excitatory connections. An activation function controls the amplitude of the output. For example, a suitable range of output is generally between 0 and 1, but in many cases it could be −1 and 1. The activation functions can be further classified into:

The Threshold Function (Eq. 4.1) accept a value of 0 if the summed input is less than a certain threshold value (v), and the value 1 if the summed input is greater than or equal to the threshold value.

$$\varphi(v) = \begin{cases} 1 & \text{if } v \geq 0 \\ 0 & \text{if } v < 0 \end{cases} \tag{4.1}$$

The Piecewise Linear Function (Eq. 4.2) can also take on the values of 0 or 1, but intermediate values between that depending on the amplification factor in a certain region of linear operation is also possible.

$$\varphi(v) = \begin{cases} 1^- & v \geq \frac{1}{2} \\ v & -\frac{1}{2} > v > \frac{1}{2} \\ 0 & v \leq -\frac{1}{2} \end{cases} \tag{4.2}$$

The Sigmoid Function (Eq. 4.3) can range between 0 and 1, or −1 to 1 range. There are different types of sigmoid function. For example, the hyperbolic tangent function can be represented by:

$$\varphi(v) = \tanh\left(\frac{v}{2}\right) = \frac{1 - \exp(-v)}{1 + \exp(-v)} \tag{4.3}$$

The Fig. 4.1 describes the basic architecture of an Artificial Neural Network. From this model, the interval activity of the neuron can be shown to be:

$$v_k = \sum_{j=1}^{p} W_{kj} X_j \tag{4.4}$$

The output of the neuron, y_k, would therefore be the outcome of some activation function on the value of v_k. If a layer of nodes are introduced in between the input and output layers then the output of the intermediate nodes will be given by the weighted sum of the inputs after modified by the activation function. Again the intermediate layer will act as the input to the output layer. That is why the output of the neural network with intermediate layers within the input and output is represented by the weighted sum of inputs after activation.

The application of neural network for optimization, prediction, or simulations can be observed in many studies from the different field of science and

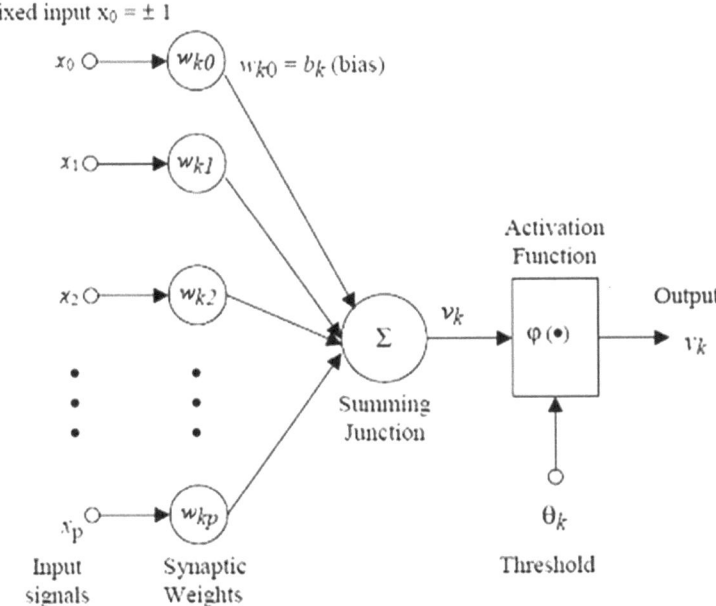

Fixed input $x_0 = \pm 1$

$w_{k0} = b_k$ (bias)

Activation Function

Output

v_k

$\varphi(\bullet)$

y_k

Σ

Summing Junction

θ_k

Threshold

Input signals

Synaptic Weights

Fig. 4.1 Showing the Network Diagram of Artificial Neural Networks

engineering. Cavalcante et al. (2013); Liu et al. (2013); Liu et al. (2013); Hu et al. (2013) etc. Interested readers may refer Doumpos and Evangelos (2013) for a collection of literatures representing all the relevant and popular application of Artificial Neural Network in decision making.

4.2 Genetic Algorithms

Genetic Algorithms (GAs) are cognitive and adaptive heuristic search algorithm developed on the evolutionary ideas of natural selection and genetic. The basic idea of GAs is intended to replicate the interactions between the cells and its chromosomes during meiotic cell division to maintain necessary developments in the natural system for evolution. The principles first set down by Charles Robert Darwin in survival of the fittest; "represent an intelligent exploitation of a random search among a defined search area to unravel a retardant".

This method of nature based algorithm was first introduced by John Holland in the year of 1960s, since then GAs have been widely studied, experimented, and applied in several fields in engineering as well as medical worlds. Several of the real-world problems involved finding optimum parameters, which could prove troublesome for traditional strategies, however, ideal for GAs.

Fathi-Moghadam et al. (2013); Jothiprakash and Arunkumar (2013); Huang et al. (2009); Penangsang et al. (2013); Deb et al. (2000); Morris et al. (1998); Krishnakumar and Goldberg (1992) are some of the examples of acutely nonlinear problems solved by GA from the field of hydropower, site selection studies, load forecasting etc. When GA is utilized to predict some parameters of neural networks, so that mapping ability of the network become more optimum it is referred as neuro-GAs.

Adib (2008); Ren (2013); Xi et al. (2013); Li et al. (2013) are some of the few who have successfully established the capacity of neuro-GAs with commendable accuracy. Each of the studies utilized the searching efficiency of GA to solve the problems of finding the optimal topology or selection of the suitable activation functions in the neural networks.

4.3 Fuzzy Logic

The concept of Fuzzy Logic (FL) was proposed by Lotfi Zadeh, a faculty member at the University of Golden State at Berkley, and conferred not as an efficient methodology, but as some way of processing information by allowing partial set membership instead of crisp set membership or nonmembership. This new avenue to set theory was not applied to control systems till the year of 1970 due to limitations in the computer capability at that time. Professor Zadeh adduced that people do not require to apply specific, numerical info as input, and however they are capable of highly adaptive management. If feedback controllers could be programmed to simply accept noisy, imprecise input, they would be much more effective and perhaps easier to implement. But industries in the USA does not embrace this technology at that time, The Europeans and Japanese on the other hand has accepted this theory and aggressively developed real merchandise around it.

FL needs some numerical parameters in order to control like what is considered significant error and significant rate-of-change-of-error, but exact values of those numbers are typically not important unless very responsive performance is needed in which case empirical calibration would determine them.

These values do not have to be symmetrical and might be "tweaked" once the system is operating in order to optimize performance.

Yager et al. (2013); Fedrizzi et al. (2013); Rao (2013); Chang et al. (2013); Agarwal et al. (2013) depicted the applications of FL in finding the solution of different nonlinear problems. Such applications for searching the optimal solution to a given problem with multiple solutions is now widely observable in various real life situations.

4.4 BAT Algorithm

BAT is a new but innovative algorithm where the solutions to various optimization problems can be found by mimicking the food foraging behaviors of microbats. Yang (2010) published a paper on BAT algorithm where he tried to solve optimization problems with the help of the echolocation property of bats which they use to differentiate between food source and barriers.

> ...The loudness also varies from the loudest when searching for prey and to a quieter base when homing towards the prey. The travelling range of such short pulses are typically a few metres, depending on the actual frequencies. Micro-bats can manage to avoid obstacles as small as thin human hairs.... (Yang 2010)

The algorithm utilizes the change in frequency of the edited sonar signals which the bats use to separate a food source from barriers in identification of the location of the optimal solution of the problem.

4.5 Analytical Hierarchy Process

As defined by DSS Resources "Analytical Hierarchy Process (AHP) (Saaty 1990) is an approach to decision making that requires structuring multiple choice criteria into a hierarchy, assessing the relative importance of these criteria, comparing alternatives for each criterion, and determining an overall ranking of the alternatives".

The concept of AHP was developed, by Thomas Saaty, an American Mathematician working at the University of Pittsburgh.

By organizing and assessing alternatives against a hierarchy of multifaceted objectives, Analytical hierarchy process (AHP) provides a proven, effective means to deal with complex decision making problems. AHP allows a better, easier, and more efficient identification of selection criteria, their weighting and analysis, which drastically reduces the decision cycle.

4.5.1 Advantage of the Analytical Hierarchy Process

AHP helps to acquire each subjective and objective analysis measures and provides a useful methodology for checking the consistency of the dissolution measures and alternatives suggested by the team therefore decreasing bias in higher cognitive process.

AHP allows organizations to minimize common pitfalls of decision-making process, like lack of focus, planning, participation, or ownership, that ultimately square measure costly distractions that can stop teams from making the correct choice.

4.5.2 Principle of the Analytical Hierarchy Process

AHP is very useful in case of unstructured decision-making process. During multi-criteria decision analysis AHP includes the alternatives taking into consideration the spread of multiple criteria that relies on multiple valued choice splitting the overall drawback into many evaluations of lesser importance but keeping the global call constant at the same time.

4.5.3 Steps of the Analytical Hierarchy Process

4.5.3.1 AHP Decomposing

The problem is divided into humanly manageable sub-problems by splitting it from top (more general) to bottom (more specific). The structure of AHP comprises of goals, criteria and alternative ratings.

The hierarchies are subdivided to appropriate level of detail to transform the unstructured difficulty into manageable problem.

When number of criteria are increased the importance of each criteria is diluted which is compensated by assigning a weight to each criterion.

4.5.3.2 AHP Weighing

A relative weight is assigned to the criteria with respect to its importance within the sub hierarchy. The summation of all the weights of sub-hierarchies must be equal to one. A global priority is computed to quantify the relative importance of a criteria within the entire decision making model.

4.5.3.3 AHP Evaluating

The alternatives are scored and compared with each other. A relative score of each of the alternative is calculated. Ultimately the variation in the relative scores of alternatives the better one among the considered alternatives is identified.

Zahedi (1986) had reviewed different applications of AHP in decision making. Besides Zahedi, Kazemi (2013), Klungboonkrong and Taylor (2013); Kim (2013); Cay and Uyan (2013); Maniya and Bhatt (2013) had applied AHP with greater level of accuracy in the predicted decisions.

References

A. Adib, Determining water surface elevation in tidal rivers by ANN. Proceedings of the ICE-Water Management **161**(2), 83–88 (2008)

M. Agarwal, K.B. Kanad, H. Madasu, Generalized intuitionistic fuzzy soft sets with applications in decision-making. Appl. Soft. Comput. (2013)

X.-S. Yang, A New Metaheuristic Bat-Inspired Algorithm, in Nature Inspired Cooperative Strategies for Optimization (NISCO 2010) (Eds. J.R. Gonzalez et al.), Studies in Computational Intelligence, 284 (Springer, Berlin, 2010), pp. 65–74

Y.L. Cavalcante, R.A. Hauser-Davis, A.C.F. Saraiva, I.L.S. Brandão, T.F. Oliveira, and A.M. Silveira, Metal and physico-chemical variations at a hydroelectric reservoir analyzed by multivariate analyses and artificial neural networks: Environmental management and policy/decision-making tools. Science of the Total Environment **442**, 509–514 (2013)

T. Cay, U. Mevlut, Evaluation of reallocation criteria in land consolidation studies using the Analytic Hierarchy Process (AHP). Land Use Policy **30**(1), 541–548 (2013)

Y.-H. Chang, C.-H. Yeh, Y.-W. Chang, A new method selection approach for fuzzy group multicriteria decision making. Appl. Soft. Comput. (2013)

K. Deb, A. Samir, P. Amrit, M. Tanaka, A fast elitist non-dominated sorting genetic algorithm for multi-objective optimization: NSGA-II. Lecture notes in computer science **1917**, 849–858 (2000)

M. Doumpos, G. Evangelos, Multicriteria decision aid and artificial intelligence: links, theory and applications. Wiley-Blackwell, (2013)

M.F. Moghadam, S. Haghighipour, H. Mohammad Vali Samani, Design-variable optimization of hydropower tunnels and surge tanks using a genetic algorithm. J. Water. Res. Planning. Manag. **139**(2), 200–208 (2013)

M. Fedrizzi, F. Michele, R.A.P. Marques, Consensus modelling in group decision making: A dynamical approach based on zadeh's fuzzy preferences. In On Fuzziness. (Springer, Berlin Heidelberg, 2013), pp. 165–170

X. Hu, C. Henning, H.A. Meyer, M. Kurt, J. Klaus, S. Carsten, Artificial neural networks and prostate cancer—tools for diagnosis and management. Nat. Rev. Urol. (2013)

H.Z. Huang, Q. Jian, J.Z. Ming, Genetic-algorithm-based optimal apportionment of reliability and redundancy under multiple objectives. IIE Transactions **41**(4), 287–298 (2009)

V. Jothiprakash, R. Arunkumar, Optimization of hydropower reservoir using evolutionary algorithms coupled with chaos. Water. Res. Manag. 1–17 (2013)

M. Kazemi, Prioritizing factors affecting bank customers using kano model and analytical hierarchy process. Int. J. Account. Financ. Manag. 6 (2013)

S.Y. Kim, Hybrid forecasting system based on case-based reasoning and analytic hierarchy process for cost estimation. J. Civil. Eng. Manag. **19**(1), 86–96 (2013)

P. Klungboonkrong, A.P.T. Michael, Application of knowledge-based expert system, Analytic hierarchy process and fuzzy set theory in multicriteria environmental sensitivity evaluation of the urban road network. KKU Eng. J. **25**(1), 1–20 (2013)

K. Krishnakumar, D. E. Goldberg. Control system optimization using genetic algorithms. J. Guidance. Control. Dyn. **15**(3), 735–740 (1992)

J. Li, Z. Zheng, F. Yu, and Z. Xiumei, Use of genetic-algorithm-optimized back propagation neural network and ordinary kriging for predicting the spatial distribution of groundwater quality parameter. In 2012 International Conference on Graphic and Image Processing. Int. Soc. Optics. Photonics. pp. 87684V–87684V (2013)

H.-H. Liu, T.-Y. Chen, Y.-H. Chiu, F.-H. Kuo, A comparison of three-stage DEA and artificial neural network on the operational efficiency of semi-conductor firms in Taiwan. (2013)

J.G. Liu, W. Yongchang, T. Tingting, C. Qingquan, Research and development of decision support system for regional agricultural development programming. In Computer and Computing Technologies in Agriculture VI. (Springer Berlin Heidelberg, 2013) pp. 271–281

K.D. Maniya, M.G. Bhatt, A selection of optimal electrical energy equipment using integrated multi criteria decision making methodology. Int. J. Energy. Optim. Eng. **2**(1), 101–116 (2013)

M.M. Garrett, D.S. Goodsell, R.S. Halliday, R. Huey, W.E. Hart, R.K. Belew, A.J. Olson, Automated docking using a lamarckian genetic algorithm and an empirical binding free energy function. J. Comput. Chem. **19**(14), 1639–1662 (1998)

O. Penangsang, A. Muhammad, R.S. Wibowo, S. Adi, Optimal design of photovoltaic–battery systems using interval type-2 Fuzzy Adaptive Genetic Algorithm. (2013)

R. Venkata, Applications of improved MADM methods to the decision making problems of manufacturing environment. In Decision Making in Manufacturing Environment Using Graph Theory and Fuzzy Multiple Attribute Decision Making Methods. (Springer London, 2013) pp. 41–135

C.-X. Ren, C.-B. Wang, C.-C Yin, M. Chen, S. Xu, The prediction of short-term traffic flow based on the niche genetic algorithm and BP neural network. In Proceedings of the 2012 International Conference on Information Technology and Software Engineering. (Springer Berlin Heidelberg, 2013) pp. 775–781

Xi, Jun, Y. Xue, Y. Xu, and Y. Shen, Artificial neural network modeling and optimization of ultrahigh pressure extraction of green tea polyphenols. Food Chemistry (2013)

Y.R. Ronald, Using agent importance to combat preference manipulation in group decision making. In Multicriteria and Multiagent Decision Making with Applications to Economics and Social Sciences. (Springer Berlin Heidelberg, 2013), pp. 301–313

F. Zahedi, The analytic hierarchy process—a survey of the method and its applications. Interfaces **16**(4), 96–108 (1986)

Chapter 5
Methodology

Abstract The problem of site selection for installation of hydropower plant can be solved by logical and scientific decision making. In the present problem, authors have used 20 different but related criteria to analyze the suitability of a location for installation of hydropower plants with the help of six popular nature-based algorithms (Neural Network, BAT Algorithms, Fuzzy Logic, Analytical Hierarchy Process, Neuro-Fuzzy, and Neuro-genetic.) Three alternatives from three different places (River Chenab in India, River Danube in Germany, and River Yukon in Alaska) which have various potential of hydropower generation (Very high, medium, none, respectively) was selected to demonstrate the efficiency of the algorithms in taking accord about this problem. The KAPPA Coefficient was utilized to compare the decision from the six different algorithms.

The Multi Criteria Decision Making methods are already used to select better alternatives based on various related criteria and a specific goal. Many studies (Yeboon and Nakayama 2013; Sanchez-Lozano et al. 2013; Ribeiro et al. 2013) has applied MCDA methods to select suitable options from the available alternatives which truly satisfies the desire or goal of the procedure. In this present study also, six different meta-heuristics are used to select the better alternative among the three available, which can satisfies the goal of identifying suitable locations for hydro power plant. The location must have enough resources for hydro power and minimum damages to the environment and hostilities from the local inhabitants. From the literatures it can be observed that hydropower plant is location dependant and the selection of location depends on some specific factors as mentioned in various reports, books, and journal papers.

5.1 Decision-Making Factors for HPP Location Selection

The factors which influence the site feasibility studies for hydropower plants can be enlisted as below:

M. Majumder and S. Ghosh, *Decision Making Algorithms for Hydro-Power Plant Location*, SpringerBriefs in Energy, DOI: 10.1007/978-981-4451-63-5_5,
© The Author(s) 2013

(1) Average Discharge

This factor is estimated based on the average annual flow variations that are observed in the location of interest. The average or mean flow represents the amount of discharge available in the location and based on which the generation of average power can also be determined. For the present investigation the calculation of average flow was estimated from a 10 years average monthly discharge data observed in the area of interest.

(2) Probability of Flow

The probability of flow describes the regularity of the flow within the selected location. The frequency duration curve or the variation of probability of flow throughout the years is an important factor for selection of a location for hydropower plants. In this factor, the area of flow duration curve of 10 years was calculated and included in the decision making. Greater the value higher will be the chance of selection.

(3) Annual Variations in the Difference in Water Level

The difference in water level which is converted to produce required kinetic energy for rotating the turbines is also included in the decision as the power production ability of the location depends directly on the available head or difference in water level between the upstream and downstream of the dam.

(4) Head Duration Curve

Similar to Flow Duration Curve the area of head duration curve is also which is included in the decision making of the present investigation. Larger the area greater will be the chance of selection as area of the curve is directly proportional to the regularity of head available in the location.

(5) Potential Power

The mean flow will also help to determine average capacity of power that can be produced. The factor is an important decider in any feasibility studies for installation of hydropower plants.

(6) Population Density

The amount of population per unit area required to be displaced is another important parameter which is included in the decision making. The decision of selecting the location inversely varies with the amount of population needed to be relocated if the power plant is approved for installation in the said location. The inhabitants will not only be displaced but also their source of income will be affected.

(7) Hostile Population

Inhabitants of the location are often hostile to Hydropower projects due to the need of relocation and rebuilding of their socioeconomic status in the new location. Thus the displaced people although compensated is often against such projects and many hydropower projects has been shelved due to disagreement with the local people. That is why in each feasibility study related to hydropower, the factor: hostility for the local inhabitants are always considered.

(8) Utilization Potential

The powers produced from the proposed plant will not only benefit the local population but it will be distributed to different domestic and industrial consumers situated in the nearby areas or connected through the electricity grids. If the volume of such consumers is large then installation of HPP becomes more profitable. That is why, HPP projects are generally approved when Utilization Potential of the power produced is justifiable. In the present study, the utilization potential is calculated by dividing Power that can be produced by number of consumers which will utilize the generated power.

(9) Distance from Nearest Grid

The power produced from HPP are generally fed to or sold to the regulation authority through an electricity grid. The distance of the grid from the location of power plant is proportional to the magnitude of loss in the transmission line. The transmission loss has to be compensated which in turn will increase the production cost of the plant. That is why grid locations near the power plant are preferred during the site selection of HPPs.

(10) Distance from Nearest Consumer

In case of nongrid connected power plants, the distance from nearest consumers to the location of HPP site controls the amount of transmission loss that can be incurred while transmitting the power to the nearest customer. That is why for projects nearer to the consumer will lower the transmission loss thus. Not only due to the transmission loss, the logistic requirement and availability of manpower decreases with increasing distance from the city to the plant.

(11) Slope

The ratio of the difference between highest and lowest point and distance between the same i.e., slope in a location influences the amount of power that can be produced and is included in the feasibility studies of any hydropower projects. The magnitude of potential energy and level of infrastructure required in storage and utilization of the energy is related to the magnitude and direction of the slope. Higher the slope higher will be the magnitude of potential energy and level of engineering works required to regulate the energy.

(12) Area of Forest Cover

The area of forest cover that can be disturbed is also considered during the feasibility studies of a HPP. In many large scale hydropower projects has been stalled due to the failure in acquiring the clearance from environmental departments or the hostility from the environmental activist which arises due to the requirement of uprooting of forest cover to install the plant infrastructure.

(13) Area of Agricultural Field

As agriculture is often the major source of income for the local population living near a river, area used for cultivation and has to be acquired for the project is an important factor to consider before a site is selected for HPP. The presence of multi-cropped agricultural land has often become a cause of contention in case of site selection of HPP projects and has induced many projects to be shelved due to the hostility from the cultivators.

(14) Area of Water Body

The presence of water bodies like lakes, ponds etc. in the proximity to the proposed location of the power plant also influence the selection procedure of the site for installation of HPP. Such water bodies are also a source of income for the local population. During the periods of water scarcity, the same water bodies act as the source of water for domestic consumption as well as for the sustenance of the agricultural activities. If installation of a project in a specific location requires filling up of large number of water bodies, then such areas are generally avoided.

(15) Presence of Wildlife

HPPs are generally selected in areas where concentration of wild animals and plants are minimal as installation of such projects destroys the habitat and also displaced them at a location in proximity to the human population which often produces instances of human–animal conflicts. Also the destruction of habitat impacts the population of the wild animals severely and even extinction of some species is reported in areas where habitats have been destroyed for installation of power plant or other large scale projects.

(16) Presence of Endangered Species

Similar to presence of wild life, if the proposed location which is being considered for HPP, is found to be the habitat of endangered flora or faunas, then such locations are generally avoided for HPPs. That is why; presence of high number of endangered species always reduces the chance of a location to be selected for hydropower projects.

(17) Tourism Potential

Installation of HPPs with storage creates a chance of tourism based on the surroundings and the reservoir itself. The natural landscapes, forest cover, etc. increase the potential of tourism of the site. If a location has a tourism potential which is beneficial for both HPP and the local population as it creates a new source of income and can compensate the damages induced due to the installation of the HPP, then the probability of selection becomes aggravated.

(18) Turbulence

The turbulence in the water body can become a factor for causing physical damages in the prime movers. The presence of turbulence in the flowing water can cause vibrations which can produce stress for the hydraulic structures like gates, penstocks etc. That is why; greater the turbulence in the water lesser is the chance of selection of the site for HPP.

(19) Amount of Sedimentation

Sedimentation in a river will reduce the depth and also suspended sediment can damage the hydraulic structures and turbines of the hydropower plants developed on such rivers. That is why rivers or channels with high amount of sediments both suspended and deposited are avoided for HPPs. In Tripura's (North East India), Gumati Reservoir deposition of sediments in the reservoir as well as the river has reduced the head that was utilized to generate hydropower. The turbines installed in the plant are of fixed head type. The high amount of sedimentation in the reservoir is now making it difficult to acquire the head required to rotate the turbine. As a result the hydropower plant is now recommended for closure.

(20) Water Quality

The water quality of the flowing water is also an important issue. Water with high amount of hardness can corrode the blades of the turbine. Suspended solids can cause physical damages to the prime movers. Also increased level of salinity can reduce the life span of the turbines. That is why the quality of river water is also considered before a river is selected for HPPs.

Along with the above factors the navigation of the fish within the rivers, soil stability of the river bed, and presence of insurgence in the command area of the proposed hydropower plant are three issues which are not considered in the present study. The navigability of fish is now compensated by fish ladders which ensure a safe path for the fish while crossing the turbines.

Again soil stability can be represented by the quality of sediments deposited in the river bed or reservoir. The concentration of turbidity also represents the fragility of the river bed and is often shows the level of instability in the soil layers within the river bed. That is why higher the turbidity lower will be the soil stability. Again if age of the sediments is old then stability of the soil will be more.

The presence of insurgency is implicitly considered in the Presence of Hostile Population factor. As insurgents are generally hostile to such projects which will help in the development of the local people the inclusion of the insurgency factor within the Hostile Population parameter is justifiable.

5.2 Nature-Based Algorithm

Nature-Based Algorithms (Jose and Talbi 2013) are nowadays popular for solving various optimization and decision-making problems. The present study employed six different nature-based algorithms to solve the present problems of selecting a location for HPP.

(1) Neuro-genetic

This algorithm is a hybridization of neural network and genetic algorithm. The genetic algorithm (Vasant et al. 2013) is mainly used for identification of the ideal network topology of the neural networks which can be applied to solve the present problem.

In case of the present investigation all the variables were at first encoded into the eleven point scale of importance. A combinatorial data matrix was developed where each and every category of importance possible within the decision variables were compared. The impact of the combinations of importance of the

Table 5.1 Table showing the scale of importance used in the investigation to represent the importance of one factor over the other

Category of importance	Symbol	Numerical counterpart (If factor is coherent to study objective)	Numerical counterpart (If factor is noncoherent to study objective)
Excessively high importance	XX	11	1
Extremely high importance	X	10	2
Very high importance	VH	9	3
High importance	H	8	4
Semi high importance	SH	7	5
Neither high neither low importance	M	6	6
Semi low importance	SL	5	7
Low importance	L	4	8
Very low importance	VL	3	9
Extremely low importance	LX	2	10
Excessively low importance	LXX	1	11

deciders was represented by Eq. 5.1 which is the ratio of coherent and noncoherent variables (to study objective) among the deciders considered in the study.

$$D = (\text{Factors Coherent to Objective/Factors Non-Coherent to Objective}) \quad (5.1)$$

The scale of importance is then converted to its numerical counterpart according to Table 5.1. The Eq. 5.1 is then normalized, ranked in ascending order to according the magnitude achieved by each row of combinations. The ranked values of the output function are then categorized into the same scale of importance but following the given logic.

Start

If

Rank of Output Function <10 % of Worst Rank Then Categorize in Excessively High Importance

Rank of Output Function <15 % of Worst Rank Then Categorize in Extremely High Importance

Rank of Output Function <25 % of Worst Rank Then Categorize in Very High Importance

Rank of Output Function <35 % of Worst Rank Then Categorize in High Importance

Rank of Output Function <45 % of Worst Rank Then Categorize in Semi High Importance

Rank of Output Function <50 % of Worst Rank Then Categorize in Neither High Neither Low Importance

Rank of Output Function <60 % of Worst Rank Then Categorize in Semi Low Importance

Rank of Output Function <70 % of Worst Rank Then Categorize in Low Importance

Rank of Output Function <80 % of Worst Rank Then Categorize in Very Low Importance

Rank of Output Function <90 % of Worst Rank Then Categorize in Extremely Low Importance

Else Categorize in Excessively Low Importance

End

The Conjugate Gradient Descent Algorithm was selected for training the models and Genetic Algorithm was utilized to select the network topology as stated earlier.

(2) Neuro-Fuzzy

In this modeling platform the weightage of all the variables were at first determined with the help of decision-making ability of fuzzy logic. The Pair wise Comparison Matrix of the considered variables was determined according to the same scale of importance that was utilized in case of the earlier modeling algorithm. Each decider was compared with the other and assigned a scale of importance with respect to the compared variable. All the ratings according to the Scale of Importance was ranked where the Excessively High Importance was assigned a rank of one and its opposite, Excessively Low Importance was awarded a rank of 11. All the rank of the importance achieved by the row variable with respect to the column variables were then divided by the maximum or worst rank. The minimum value attained from this operation was complemented and was taken as the weightage of the row variable.

Once the weightage of the variables were determined, the same combinatorial database that was prepared for the earlier meta-heuristic where all the class of importance that a variable can be encoded into was generated so that each possible situation that may arise between the variables can be included in the database for training of the neural network which will again estimate the impact of the combinations of the decider variables on the decision. Here Eq. 5.1 is modified into Eq. 5.2:

$$\frac{\sum_{n=1}^{n} nxVn}{\sum_{n=1}^{n} nx \sum_{n=1}^{n} Vn} \tag{5.2}$$

n (1 to 20) is the weightage of the variables as determined by fuzzy logic.

The numerical counterparts of the ratings are retrieved from Table 5.1. The output is also normalized and encoded into same class of importance according to the logic described in the earlier section.

Table 5.2 Table showing the scale of Importance used in the investigation to represent the importance of one factor over the other

Category of importance	Symbol
Excessively high importance	XX
Extremely high importance	X
Very high importance	VH
High importance	H
Semi high importance	SH
Neither high neither low importance	M
Semi low importance	SL
Low importance	L
Very low importance	VL
Extremely low importance	LX
Excessively low importance	LXX

The same training algorithm, Conjugate Gradient Descent was utilized to train the model but GA was not applied to identify the network topology, instead, trial, and error method was endowed to find the number of hidden layers optimum for the present type of problem (Tables 5.3 and 5.4).

(3) Fuzzy Logic

The fuzzy logic was applied to determine the weightage. The product of weightage and the rating for the input variables; which are proportional to the study objective; was divided by the product of the weightage and rating of the variables; which are incoherent to the present objective; to find the overall rating of an alternative (Tables 5.5 and 5.6) (Felipe et al. 2013).

Table 5.3 Table showing the attributes of Neuro-genetic model

Model output variable	Ration of product of coherent to noncoherent decision making variables
Weightage determined by	None
Type of neural network	Feed forward
Topology selection	Genetic algorithm
Activation function	Sinusoidal
Number and type of inputs	22-Categorical
Number and type of outputs	1-Categorical
Training	Conjugate gradient descent
Stop training conditions	
Correct classification rate required	99 %
Maximum generalization loss allowed	50 %
Validation of model	
Performance metrics for training performance analysis	Correct classification rate
Performance metrics for training performance analysis	Correct classification rate

Table 5.4 Table showing the attributes of Neuro-Fuzzy model

Model output variable	Ration of weighted product of coherent to noncoherent decision making variables
Weightage determined by	Fuzzy logic
Type of neural network	Feed forward
Topology selection	Trial and error
Activation function	Sinusoidal
Number and type of inputs	22-Categorical
Number and type of outputs	1-Categorical
Training	Conjugate gradient descent
Stop training conditions	
Correct classification rate required	99 %
Maximum generalization loss allowed	5 %
Validation of model	
Performance metrics for training performance analysis	Correct classification rate
Performance metrics for training performance analysis	Correct classification rate

(4) BAT Algorithm

BAT algorithm is commonly utilized in different optimization studies and follows the simple food foraging behavior of the microbats. But in the present case, the objective was to select the better alternative available. The methodology of BAT algorithm which is utilized in optimization studies were modified and applied to find the better alternative among many options available. No circular parameter was included and also the goal was limited to conduction of a clear and distinct decision making study.

Any microbat when it spots a source of food emits an auditory sub-sonic signal to attract other microbats toward the food location. When it identifies an optimal location of food source it increases both frequency and loudness of the auditory signal.

Similar to this logic in the present investigation, the coherent factors were taken similar to the location of good quality food source and incoherent factors were assumed to the locations which must be ignored. The weightage of all the factors were varied randomly in between 0 and 1 and product of all the good and bad source factors were separately determined. As the weights were randomly varied the difference between the two products also changed. The weightage at which the difference between the two kinds of factors becomes largest was collected as the weightage at which the most optimum and distinct decision can be possible from the objective equation described in Eq. 5.1 (Tables 5.7 and 5.8).

(5) Analytical Hierarchy Process

In this method, all the variable are compared with each other to assign a rating which indicates the importance of the variable with respect to the other variable. The rating must be either in odd or even numbers but if the importance is not clear enough both odd and even numbers can be utilized.

Table 5.5 Table showing the pair wise comparison matrix to find the Fuzzy rate of importance for the factors for Neuro-Fuzzy and Fuzzy logic method of decision making

	AQ	aPAQ	AH	aPAH	WQI	S	E	T	P	%P	Pd	F	C	W	A POP	ENDA	TOU	P	U	Dg	dc	s
AQ	NINU	AVG	H	H	EX	EX	EX	EX	EX	EX	EXX	EXX	EXX	EXX	EXX	EXX	EXX	SH	H	H	H	H
aPAQ	AVG	NINU	H	H	EX	EX	EX	EX	EX	EX	EXX	EXX	EXX	EXX	EXX	EXX	EXX	SH	H	H	H	H
AH	L	L	NINU	AVG	VH	VH	VH	VH	VH	VH	EX	EX	EX	EX	EX	EX	EX	AVG	H	H	H	H
aPAH	L	L	AVG	NINU	VH	VH	VH	VH	VH	VH	EX	EX	EX	EX	EX	EX	EX	AVG	H	H	H	H
WQI	LX	LX	VL	VL	NINU	AVG	SL	SL	AVG	SL	SL	SL	SL	SL	SL	H	H	L	L	L	L	L
S	LX	LX	VL	VL	AVG	NINU	AVG	SL	AVG	L	SL	H	H	H	H	SH	VH	L	L	L	L	SH
E	LX	LX	VL	VL	SH	AVG	NINU	AVG	L	L	L	AVG	SH	H	H	H	VH	L	L	L	L	SH
T	LX	LX	VL	VL	SH	SH	AVG	NINU	AVG	AVG	L	H	H	H	H	H	EX	AVG	AVG	SH	SH	SH
P	LX	LX	VL	VL	SL	SL	H	AVG	NINU	L	VL	L	L	L	L	L	H	VL	VL	L	L	L
%P	LX	LX	VL	VL	SH	SH	H	AVG	H	NINU	H	VH	VH	VH	VH	H	EX	SH	SH	SH	SH	L
Pd	LX	LXX	LX	LX	SH	SL	H	AVG	VH	H	NINU	VH	VH	VH	SH	AVG	H	H	H	H	H	L
F	LXX	LXX	LX	LX	SH	L	AVG	L	H	VL	VL	NINU	SL	AVG	SH	AVG	L	L	L	SL	SL	L
C	LXX	LXX	LX	LX	SH	L	SH	L	H	VL	VL	SL	NINU	AVG	SH	SL	SL	SL	SH	SL	SL	H
W	LXX	LXX	LX	LX	SH	L	H	L	H	VL	VL	SL	AVG	NINU	SH	SL	SL	SL	L	SL	SL	H
A POP	LXX	LXX	LX	LX	AVG	L	SL	L	H	VH	VH	AVG	AVG	AVG	NINU	SL	H	H	SL	L	L	H
ENDA	LXX	LXX	LX	LX	L	VL	SL	LX	L	LX	L	H	SH	SH	SH	NINU	H	SL	SL	SL	SL	L
TOU	LXX	LXX	LX	LX	H	H	VL	AVG	H	SL	L	H	SH	SH	H	L	NINU	LX	LX	VL	VL	VH
P	SL	SL	AVG	AVG	H	H	H	AVG	L	SL	L	H	SH	SH	H	SH	EX	NINU	AVG	SH	SH	VH
U	L	L	L	L	H	H	H	AVG	VH	SL	L	H	SH	SH	H	SH	EX	AVG	NINU	SH	SH	VH
Dg	L	L	L	L	H	H	H	SL	H	SL	L	H	SH	SH	H	SH	VH	SL	SL	NINU	AVG	AVG
dc	L	L	L	L	H	H	H	SL	H	SL	H	H	SH	SH	H	SH	VH	SL	SL	AVG	NINU	AVG
s	L	L	L	L	SL	SH	SH	SL	H	L	L	L	H	H	AVG	H	VH	VH	VH	AVG	AVG	NINU

Table 5.6 Table showing the crisp values of the Fuzzy ratings of the factors and the worst rank achieved by the variables for the Fuzzy and Neuro-Fuzzy method of decision making

	AQ	aPAQ	AH	aPAH	WQI	S	E	T	P	%P	Pd	F	C	W	A POP	ENDA	TOU	P	U	Dg	dc	s	Worst rank
AQ	1	6	4	4	2	3	4	5	6	7	1	1	1	1	1	1	1	5	4	4	4	4	7
aPAQ	6	1	4	4	2	3	4	5	6	7	1	1	1	1	1	1	1	5	4	4	4	4	7
AH	8	8	1	6	3	3	3	3	3	3	2	3	4	5	6	7	8	6	4	4	4	4	8
aPAH	8	8	6	1	3	3	3	3	3	3	2	3	4	5	6	7	8	6	4	4	4	4	8
WQI	10	10	9	9	1	1	7	7	6	7	7	7	7	7	7	1	4	8	8	8	8	8	10
S	10	10	9	9	6	1	6	6	6	7	7	4	4	4	4	4	3	8	8	8	8	8	10
E	10	10	9	9	5	1	1	6	8	8	8	6	5	5	5	5	3	8	8	8	8	5	10
T	10	10	9	9	5	5	6	1	6	6	6	4	4	4	4	4	1	6	6	5	5	5	10
P	10	10	9	9	7	7	4	6	1	8	9	8	8	8	8	8	4	9	9	8	8	8	10
%P	10	10	9	9	5	5	4	6	4	1	4	3	3	3	3	4	1	5	5	5	5	8	10
Pd	11	11	10	10	5	5	4	6	3	8	1	3	3	3	3	4	4	4	5	4	4	8	11
F	11	11	10	10	5	8	6	8	4	9	9	1	5	5	5	6	8	8	8	8	8	8	11
C	11	11	10	10	5	8	6	8	4	9	9	7	1	6	5	6	7	7	7	7	7	8	11
W	11	11	10	10	5	8	7	8	4	9	9	7	6	1	6	7	7	7	7	7	7	8	11
A POP	11	11	10	10	5	8	7	8	4	9	9	7	7	6	1	7	8	8	7	8	8	8	11
ENDA	11	11	10	10	6	8	7	8	4	3	3	6	6	5	5	1	4	7	7	7	7	8	11
TOU	11	11	10	10	8	9	9	10	8	10	8	4	5	5	4	8	1	10	10	9	9	8	11
P	7	7	6	6	4	4	4	6	3	7	8	4	5	5	4	5	2	1	6	5	5	3	8
U	8	8	8	8	4	4	4	6	3	7	8	4	5	5	4	5	2	6	1	5	5	3	8
Dg	8	8	8	8	4	4	4	7	4	7	8	4	5	5	4	5	3	7	7	1	6	1	8
dc	8	8	8	8	4	4	4	7	4	7	8	4	5	5	4	5	3	7	7	6	1	1	8
s	8	8	8	8	4	7	7	7	4	4	4	4	4	4	4	4	4	9	9	1	1	1	9

Table 5.7 Table showing the description of the BAT algorithm

Objective	To find optimal location of food source	To find the optimal combination of weightage so that the ideal alternative can be selected
Procedure adopted	The microbats try to identify and differentiate in between the location of a good source and bad source food.	The weightage of the coherent and noncoherent variables were updated and difference between the weighted product of two type of variables are identified so that a pattern can be found when the set of weightage will maximum difference between the two types of variables which will in turn help to determine the weightage at which the difference between coherent and non-coherent variables is maximum. The weightages can be utilized to separate the better alternative among the many other for a given decision making problem.
Parameter of identifying the optimal location	Frequency and loudness of Sonar Signal emitted from the BAT which has located the food source	Weightages of the criteria which are randomized
Fitness function	The frequency and loudness of sonar signal emitted from the micro-bat which has identified the food source. The frequency and loudness changes according to the quality of food source.	Difference between the weighted product of coherent and non coherent variables

After each decider has been assigned a rating of importance with respect to the other it is divided with the worst rate achieved by the variable with respect to other variable. After this operation, the average of the result of the division is calculated and selected as the weightage of that variable.

Then all the alternatives are rated with respect to the other options based on each of the deciders. The alternatives are assigned a rating of importance by either even or odd number similar to the rating mechanism of the deciders. This ratings are also divided with the worst rating and the average of all the ratings achieved by an alternative with respect to the other options are taken as the weightage of that option with respect to the decider based on which the rate of importance is assigned.

The same mechanism is repeated for all the deciders and ultimately a matrix is prepared where alternatives and their rating achieved with respect to the deciders are given in the rows and the deciders are placed in the columns. The weightage of the alternatives and weightage of the criteria is multiplied and summed up to estimate the resultant weightage of each of the alternatives. The

Table 5.8 Table showing some of the iterations performed under BAT algorithms to find the set of weightage for which the difference between coherent and noncoherent will be maximum

Criteria	Iterations						
Average Flow	X	VH	H	SH	N	SL	XX
Random weightage	0.957116	0.843145	0.839705	0.862909	0.83353	0.96879	0.986753
Numerical value of the rating of importance	10	9	8	7	6	5	11
Annual probability of average flow	X	VH	H	SH	N	SL	XX
Random weightage	0.989641	0.996913	0.999016	0.997751	0.997132	0.992924	0.994869
Numerical value of the rating of importance	10	9	8	7	6	5	11
Average head	X	VH	H	SH	N	SL	XX
Random weightage	0.932532	0.962724	0.973246	0.936104	0.83894	0.918282	0.884691
Numerical value of the rating of importance	10	9	8	7	6	5	11
Annual probability of average head	X	VH	H	SH	N	SL	XX
Random weightage	0.954531	0.927424	0.880242	0.91424	0.956106	0.900237	0.922707
Numerical value of the rating of importance	10	9	8	7	6	5	11
Water quality index	X	VH	H	SH	N	SL	XX
Random weightage	0.82921	0.806341	0.794795	0.759598	0.857889	0.932087	0.863547
Numerical value of the rating of importance	10	9	8	7	6	5	11
Potential of tourism	X	VH	H	SH	N	SL	XX
Random weightage	0.963639	0.976598	0.971427	0.958583	0.993162	0.983815	0.962944
Numerical value of the rating of importance	10	9	8	7	6	5	11
Power potential	X	XX	X	XX	X	XX	XX
Random weightage	0.96441	0.979049	0.970701	0.948826	0.976628	0.946243	0.991955
Numerical value of the rating of importance	10	11	10	11	10	11	11
Utilization potential	X	VH	H	SH	N	SL	XX
Random weightage	0.999255	0.999641	0.999659	0.998352	0.999027	0.99938	0.99827
Numerical value of the rating of importance	10	9	8	7	6	5	11
Distance to nearest grid	X	XX	X	XX	X	XX	XX

(continued)

Table 5.8 (continued)

Criteria	Iterations						
Average Flow	X	VH	H	SH	N	SL	XX
Random weightage	0.977973	0.998112	0.978131	0.959143	0.972003	0.953432	0.957205
Numerical value of the rating of importance	10	11	10	11	10	11	11
Distance to nearest consumer	XX	XX	XX	XX	XX	XX	XX
Random weightage	0.989993	0.981071	0.996392	0.97911	0.973135	0.999307	0.970459
Numerical value of the rating of importance	11	11	11	11	11	11	11
Slope	XX	XX	XX	XX	XX	XX	XX
Random weightage	0.999444	0.999282	0.998748	0.999317	0.999873	0.999046	0.998812
Numerical value of the rating of importance	11	11	11	11	11	11	11
Sedimentation	LX	L	SL	N	SH	H	LXX
Random weightage	0.769628	0.844308	0.951623	0.762051	0.849569	0.918715	0.753156
Numerical value of the rating of importance	2	4	5	6	7	8	1
Bank erosion	X	X	X	X	X	X	X
Random weightage	0.734376	0.534165	0.720912	0.964879	0.606853	0.843201	0.423249
Numerical value of the rating of importance	10	10	10	10	10	10	10
Turbulence	LX	L	SL	N	SH	H	LXX
Random weightage	0.888698	0.982293	0.725881	0.939911	0.617291	0.758713	0.744029
Numerical value of the rating of importance	2	4	5	6	7	8	1
Population concentration	LXX	LXX	LXX	LXX	LXX	LXX	LXX
Random weightage	0.85126	0.568653	0.902367	0.644238	0.68916	0.668932	0.807734
Numerical value of the rating of importance	1	1	1	1	1	1	1
Percentage hostile	X	VH	H	SH	N	SL	XX
Random weightage	0.848227	0.931049	0.783533	0.809106	0.830647	0.885672	0.90204
Numerical value of the rating of importance	10	9	8	7	6	5	11
Estimated displacement of population	X	VH	H	SH	N	SL	XX
Random weightage	0.788784	0.441368	0.337592	0.355564	0.537854	0.771611	0.319526

(continued)

Table 5.8 (continued)

Criteria	Iterations						
Average Flow	X	VH	H	SH	N	SL	XX
Numerical value of the rating of importance	10	9	8	7	6	5	11
Estimated area of affected forest	X	X	X	X	X	X	X
Random weightage	0.693105	0.580749	0.713814	0.897708	0.381524	0.403893	0.515321
Numerical value of the rating of importance	10	10	10	10	10	10	10
Estimated area of affected crop fields	X	XX	X	XX	X	SL	XX
random weightage	0.938058	0.983227	0.943759	0.998308	0.985255	0.968098	0.970724
Numerical value of the rating of importance	10	11	10	11	10	5	11
Estimated area of effected waterbodies	X	VH	H	SH	N	SL	XX
Random weightage	0.916458	0.789125	0.627318	0.753556	0.599105	0.866367	0.951976
Numerical value of the rating of importance	10	9	8	7	6	5	11
Estimated concentration of effected animal population	X	VH	H	SH	XX	X	XX
Random weightage	0.8984	0.618302	0.825404	0.956903	0.812182	0.998773	0.602154
Numerical value of the rating of importance	10	9	8	7	11	10	11
Concentration of endangered/rare animals	X	VH	H	SH	XX	X	XX
Random weightage	0.979765	0.95875	0.997054	0.942271	0.95771	0.95617	0.972127
Numerical value of the rating of importance	10	9	8	7	11	10	11
Source	107.5669	102.1477	93.09958	87.74312	80.04412	76.3559	115.8543
No source	72.13964	63.48704	61.64193	67.18507	61.96771	63.56747	63.59464
Fitness (difference between source and nonsource)	35.42725	38.66069	31.45765	20.55805	18.07641	12.78843	52.2597

resultant weightage is normalized and compared with the same of the other alternatives to identify the maximum weightage achieved by an alternative which will be selected as the better alternative among the options considered according to the AHP method of decision making (Tables 5.9, 5.10, and 5.11).

(6) Neural Networks

The advantage of neural networks is utilized to estimate the decision for selection of better site location for a HydroPower Plant among the three alternatives considered. A combinatorial data matrix was prepared considering each and every possible combination of the input and output variables and the eleven ratings that can be assigned to them to represent their degree of importance over the other options.

The output of the decision was calculated with the help of Eq. 5.1 by converting the ratings into its numerical counterpart and multiplying the values of the rating of the deciders which are coherent with the objective and dividing the same with the product of all the noncoherent deciders. In this method no weightage representing the importance of each of the deciders are incorporated. In this case also the neural network is trained with Conjugate Gradient Descent and the topology is selected with the help of trial and error.

In case of all the models the same eleven point scale of importance was used. All the parameters were rated according to their conditions in the given alternatives. The algorithms were applied to determine the weightage of all the factors. (Table 5.12).

The efficacy of all the six meta-heuristic algorithms was verified with the help of three case studies in different places of the world. In one of the locations a large scale hydropower plant is already installed and running successfully. In the second location a large scale hydropower plant is planned but not installed till date. The third location was selected in such a way that it has no potential to be utilized as a HP location.

The efficiency of the six nature-based algorithms is compared by applying them to select the better option out of the three locations considered for the present study and as described above.

The three different locations considered in the present study are

1. River Chenab in India: A large scale HPP is running in the Baglihar Dam
2. Upstream of River Danube in Germany: HPP planned and sanctioned waiting for installation
3. Downstream of River Yukon in Alaska: No HPP is possible in this location

The differences in geo-morphological, ecological, and socio-economical characteristics of the three locations were utilized to verify the performance of the six different meta-heuristics applied in the decision making for the present investigation of selecting the optimal location for installation of HPPs (Table 5.13).

Table 5.9 Table shows the ratings achieved by each of the criteria with respect to the other as per the scale of importance for the method of AHP

Criteria	AQ	aPAQ	AH	aPAH	WQI	S	E	T	P	%P	Pd	F	C	W	A POP	ENDA	TOU	P	U	Dg	dc	s
AQ	NINU	N	L	L	X	X	X	X	X	X	XX	XX	XX	XX	XX	XX	XX	SH	H	H	H	H
aPAQ	N	NINU	L	L	X	X	X	X	X	X	XX	XX	XX	XX	XX	XX	XX	SH	H	H	H	H
AH	L	L	NINU	N	VH	VH	VH	VH	VH	VH	X	X	X	X	X	X	X	N	H	H	H	H
aPAH	L	L	N	NINU	VH	VH	VH	VH	VH	VH	X	X	X	X	X	X	H	N	H	H	H	L
WQI	LX	LX	VL	VL	NINU	N	SL	SL	N	SL	SL	SL	SL	N	N	SL	SL	L	L	L	L	L
S	LX	LX	VL	VL	N	NINU	N	SL	N	SL	SL	H	H	H	H	SH	VH	H	H	H	H	SH
E	LX	LX	VL	VL	SH	N	NINU	N	L	L	L	N	SH	SH	SH	SH	VH	H	H	H	L	SH
T	LX	LX	VL	VL	SH	SH	NINU	NINU	N	N	VL	N	SH	N	SL	SL	L	H	N	SH	SH	SH
P	LX	LX	VL	VL	SH	SL	H	N	NINU	L	VL	L	L	N	SL	L	NINU	L	VL	L	L	L
%P	LX	LX	VL	VL	SH	SH	H	N	N	NINU	H	VH	VH	VH	VH	H	X	SL	SL	SH	SH	L
Pd	LXX	LXX	LX	LX	SH	SH	H	L	H	NINU	H	VH	VH	VH	H	SH	H	L	H	H	H	L
F	LXX	LXX	LX	LX	SH	SH	L	L	H	L	NINU	SL	VH	VH	SH	N	L	L	L	L	L	L
C	LXX	LXX	LX	LX	SH	SH	L	L	H	VL	SL	NINU	N	N	N	SL	SL	SL	SL	SL	SL	L
W	LXX	LXX	LX	LX	SH	SH	L	L	H	VL	SL	N	NINU	N	N	SL	SL	SL	SL	SL	SL	L
A POP	LXX	LXX	LX	LX	SH	SH	L	L	H	VL	SL	N	NINU	N	NINU	SL	L	L	L	L	L	L
ENDA	LXX	LXX	LX	LX	N	H	L	L	H	VH	VH	N	N	SL	SL	NINU	H	SL	SH	SH	SH	L
TOU	LXX	LXX	LX	LX	L	H	LX	LX	L	LX	L	SH	SH	SH	H	L	NINU	LX	LX	VL	VL	VH
P	SL	SL	N	N	H	H	H	N	VH	SL	L	SH	SH	SH	H	SH	X	NINU	N	SH	SH	VH
U	L	L	L	L	H	H	H	N	VH	SL	L	SH	SH	SH	H	SH	X	NINU	NINU	SH	SH	VH
Dg	L	L	L	L	H	H	H	SL	H	SL	L	SH	SH	NINU	H	SH	VH	SL	NINU	NINU	N	AVG
dc	L	L	L	L	H	H	H	SL	H	SL	L	SH	SL	N	H	SH	VH	SL	SL	N	NINU	AVG
s	H	H	H	H	L	SH	SH	SH	L	L	L	L	L	L	L	L	L	VH	VH	AVG	AVG	NINU

Table 5.10 Table shows the pair wise comparison matrix of the criteria considered and its numerical ratings of importance for AHP method of decision making

Criteria	AQ	aPAQ	AH	aPAH	WQI	S	E	T	P	%P	Pd	F	C	W	A POP	ENDA	TOU	P	U	Dg	dc	s	Average	Normalized average
AQ	1	1.000	5.000	5.000	9.000	9.000	9.000	9.000	9.000	9.000	11.000	11.000	11.000	11.000	11.000	11.000	11.000	3.000	5.000	5.000	5.000	5.000	7.500	0.125
aPAQ	1.000	1	5.000	5.000	9.000	9.000	9.000	9.000	9.000	9.000	11.000	11.000	11.000	11.000	11.000	11.000	11.000	3.000	5.000	5.000	5.000	5.000	7.500	0.125
AH	0.200	0.200	1	1.000	7.000	7.000	7.000	7.000	7.000	3.000	9.000	9.000	9.000	9.000	9.000	9.000	9.000	1.000	5.000	5.000	5.000	5.000	5.791	0.096
aPAH	0.200	0.200	1.000	1	7.000	7.000	7.000	7.000	7.000	3.000	9.000	9.000	9.000	9.000	9.000	9.000	9.000	1.000	5.000	5.000	5.000	5.000	5.791	0.096
WQI	0.111	0.111	0.143	0.143	1	1.000	1.000	0.333	1.000	0.333	0.333	5.000	5.000	3.000	5.000	5.000	5.000	0.200	0.200	0.200	0.200	0.200	0.553	0.009
S	0.111	0.111	0.143	0.143	1.000	1	1.000	1.000	1.000	0.333	0.333	5.000	5.000	5.000	5.000	5.000	7.000	0.200	0.200	0.200	0.200	3.000	1.832	0.030
E	0.111	0.111	0.143	0.143	1.000	1.000	1	1.000	0.200	0.200	0.200	1.000	3.000	1.000	1.000	1.000	5.000	0.143	0.200	0.200	0.200	3.000	1.359	0.023
T	0.111	0.111	0.143	0.143	3.000	1.000	1.000	1	0.200	1.000	0.143	0.333	1.000	1.000	0.111	1.000	0.200	0.111	1.000	0.143	0.143	3.000	2.523	0.042
P	0.111	0.111	0.143	0.143	1.000	1.000	5.000	5.000	1	1.000	1.000	5.000	5.000	5.000	5.000	5.000	5.000	0.200	7.000	5.000	5.000	0.200	0.655	0.011
%P	0.111	0.111	0.143	0.143	0.333	1.000	1.000	1.000	1.000	1	5.000	7.000	7.000	7.000	7.000	7.000	7.000	0.333	0.333	0.333	0.333	0.200	3.487	0.058
Pd	0.111	0.111	0.143	0.143	3.000	3.000	3.000	1.000	5.000	1	1	7.000	7.000	7.000	7.000	7.000	5.000	0.200	0.200	0.200	0.200	0.200	3.537	0.059
F	0.091	0.091	0.111	0.111	3.000	3.000	3.000	3.000	5.000	5.000	1	1	3.000	1.000	1.000	1.000	0.200	5.000	5.000	5.000	5.000	0.200	0.968	0.016
C	0.091	0.091	0.111	0.111	3.000	3.000	3.000	3.000	5.000	5.000	5.000	1.000	1	1.000	1.000	1.000	0.333	0.200	0.333	0.333	0.333	0.200	0.786	0.013
W	0.091	0.091	0.111	0.111	3.000	3.000	3.000	3.000	5.000	5.000	5.000	1.000	1	1	1.000	1.000	0.333	0.200	0.333	0.333	0.333	0.200	0.634	0.011
A POP	0.091	0.091	0.111	0.111	1.000	1.000	1.000	0.333	1.000	0.333	0.143	1.000	1.000	1	1	1.000	0.200	3.000	5.000	5.000	5.000	0.200	0.574	0.010
ENDA	0.091	0.091	0.111	0.111	1.000	1.000	1.000	0.333	1.000	7.000	0.200	1.000	1.000	1.000	1	1	5.000	0.111	0.333	0.333	0.333	0.200	1.621	0.027
TOU	0.091	0.091	0.111	0.111	0.200	5.000	5.000	0.143	5.000	0.111	0.333	1.000	3.000	5.000	5.000	1	1	0.111	0.111	0.143	0.143	0.200	0.837	0.014
P	0.333	0.333	1.000	1.000	5.000	5.000	5.000	1.000	7.000	0.333	0.200	5.000	3.000	3.000	5.000	3.000	9.000	1	1.000	3.000	3.000	7.000	3.100	0.052
U	0.200	0.200	0.200	0.200	5.000	5.000	5.000	1.000	7.000	0.333	0.200	5.000	3.000	3.000	5.000	3.000	9.000	1.000	1	3.000	3.000	7.000	3.015	0.050
Dg	0.200	0.200	0.200	0.200	5.000	5.000	5.000	0.333	5.000	0.333	0.200	5.000	3.000	3.000	5.000	3.000	7.000	0.333	0.333	1	1.000	11.000	2.742	0.046
dc	0.200	0.200	0.200	0.200	5.000	5.000	5.000	0.333	5.000	0.333	0.200	5.000	3.000	3.000	5.000	3.000	7.000	0.333	0.333	1.000	1	11.000	2.288	0.038
s	5.000	5.000	5.000	5.000	0.200	3.000	3.000	3.000	0.200	0.200	0.200	0.200	0.200	0.200	0.200	0.200	0.200	0.200	7.000	7.000	11.000	1	3.045	0.051

Table 5.11 Table showing the determination of weightage for the alternatives w. r. t each of the criteria

AQ	A1	A2	A3	Average	Normalized average
A1	0	5	0.33	1.777	0.388
A2	0.20	0	0.20	0.133	0.029
A3	3	5	0	2.666	0.582
APAQ	A1	A2	A3	Average	Normalized average
A1	0.000	5.000	0.333	1.777	0.388
A2	0.200	0.000	0.200	0.133	0.029
A3	3.000	5.000	0.000	2.666	0.582
AH	A1	A2	A3	Average	Normalized average
A1	0.000	3.000	7.000	3.333	0.637
A2	0.333	0.000	5.000	1.777	0.340
A3	0.143	0.200	0.000	0.114	0.021
APAH	A1	A2	A3	Average	Normalized average
A1	0.000	3.000	7.000	3.333	0.637
A2	0.333	0.000	5.000	1.777	0.340
A3	0.143	0.200	0.000	0.114	0.021
WQI	A1	A2	A3	Average	Normalized average
A1	0.000	0.333	3.000	1.111	0.280
A2	3.000	0.000	5.000	2.666	0.674
A3	0.333	0.200	0.000	0.177	0.044
Pd	A1	A2	A3	Average	Normalized average
A1	0.000	3.000	5.000	2.666	0.582
A2	0.333	0.000	0.200	0.177	0.038
A3	0.200	5.000	0.000	1.733	0.378640777
F	A1	A2	A3	Average	Normalized average
A1	0.000	3.000	5.000	2.666	0.674
A2	0.333	0.000	3.000	1.111	0.280

s	A1	A2	A3	Average	Normalized average
A1	0.000	3.000	5.000	2.666	0.674
A2	0.333	0.000	3.000	1.111	0.280
A3	0.200	0.333	0.000	0.177	0.044
E	A1	A2	A3	Average	Normalized average
A1	0.000	3.000	5.000	2.666	0.674
A2	0.333	0.000	3.000	1.111	0.280
A3	0.200	0.333	0.000	0.177	0.044
T	A1	A2	A3	Average	Normalized average
A1	0.000	3.000	7.000	3.333	0.637
A2	0.333	0.000	5.000	1.777	0.340
A3	0.143	0.200	0.000	0.114	0.021
P	A1	A2	A3	Average	Normalized average
A1	0.000	3.000	0.333	1.111	0.280
A2	0.333	0.000	0.200	0.177	0.044
A3	3.000	5.000	0.000	2.666	0.674
%P	A1	A2	A3	Average	Normalized average
A1	0.000	7.000	5.000	4	0.765
A2	0.143	0.000	0.333	0.158	0.030
A3	0.200	3.000	0.000	1.066	0.204
ENDA	A1	A2	A3	Average	Normalized average
A1	0.000	3.000	3.000	2	0.692
A2	0.333	0.000	1.000	0.444	0.153
A3	0.333	1.000	0.000	0.444	0.153
TOU	A1	A2	A3	Average	Normalized average
A1	0.000	3.000	5.000	2.666	0.674
A2	0.333	0.000	3.000	1.111	0.280

(continued)

Table 5.11 (continued)

AQ	A1	A2	A3	Average	Normalized average
A3	0.200	0.333	0.000	0.177	0.044
C	A1	A2	A3	Average	Normalized average
A1	0.000	1.000	0.333	0.444	0.153
A2	1.000	0.000	0.333	0.444	0.153
A3	3.000	3.000	0.000	2	0.692
W	A1	A2	A3	Average	Normalized average
A1	0.000	5.000	3.000	2.666	0.674
A2	0.200	0.000	3.000	1.066	0.269
A3	0.333	0.333	0.000	0.222	0.056
A POP	A1	A2	A3	Average	Normalized average
A1	0.000	3.000	3.000	2	0.692
A2	0.333	0.000	1.000	0.444	0.153
A3	0.333	1.000	0.000	0.444	0.153
Dc	A1	A2	A3	Average	Normalized average
A1	0.000	1.000	0.333	0.444	0.153
A2	1.000	0.000	0.333	0.444	0.153
A3	3.000	3.000	0.000	2	0.692

s	A1	A2	A3	Average	Normalized average
A3	0.200	0.333	0.000	0.177	0.044
P	A1	A2	A3	Average	Normalized average
A1	0.000	5.000	3.000	2.666	0.674
A2	0.200	0.000	0.333	0.177	0.044
A3	0.333	3.000	0.000	1.111	0.280
U	A1	A2	A3	Average	Normalized average
A1	0.000	3.000	5.000	2.666	0.674
A2	0.200	0.000	3.000	1.111	0.280
A3	0.200	0.333	0.000	0.177	0.044
Dg	A1	A2	A3	Average	Normalized average
A1	0.000	1.000	0.333	0.444	0.153
A2	1.000	0.000	0.333	0.444	0.153
A3	3.000	3.000	0.000	2	0.692
s	A1	A2	A3	Average	Normalized average
A1	0.000	3.000	7.000	3.333	0.637
A2	0.333	0.000	5.000	1.777	0.340
A3	0.143	0.200	0.000	0.114	0.021

Table 5.12 Table showing the attributes of neural network model

Model output variable	Ration of coherent to noncoherent decision making variables
Weightage determined by	None
Type of neural network	Feed forward
Topology selection	Trial and error
Activation function	Sinusoidal
Number and type of inputs	22-Categorical
Number and type of outputs	1-Categorical
Training	Conjugate gradient descent
Stop training conditions	
Correct classification rate required	99%
Maximum generalization loss allowed	5%
Validation of model	
Performance metrics for training performance analysis	Correct classification rate
Performance metrics for training performance analysis	Correct classification rate

Table 5.13 Table showing the attributes of the present investigation

Objective	Location selection of SHP
Input parameters	Average Flow, Annual Probability of Average Flow, Average Head, Annual Probability of Average, Head, Water Quality Index, Sedimentation, Bank Erosion, Turbulence, Population Concentration, Percentage Hostile, Estimated Displacement of Population, Estimated Area of Affected Forest, Estimated Area of Affected Cropfields, Estimated Area of Effected Water bodies, Estimated Concentration of Effected Animal Population, Concentration of Endangered/Rare Animals, Potential of Tourism, Power Potential, Utilization Potential, Distance to Nearest Grid, Distance to Nearest Consumer, Slope.
Output parameters	Decision of Location Selection for SHP
Algorithms applied	Neural Networks, Fuzzy Logic, BAT, Analytical Hierarchy Process, Neuro-Genetic, Neuro-Fuzzy
Case studies	River Chenab in India; River Danube in Germany, and River Yukon in Alaska

References

R.F. José, E.-G. Talbi, Emergent nature inspired algorithms for multi-objective optimization. Comput. Oper. Res. (2013). doi:http://dx.doi.org/10.1016/j.cor.2013.01.020

Y. Yeboon, H. Nakayama, Generalized data envelopment analysis and computational intelligence in multiple criteria decision making. Multicrit. Decis. Aid Artif. Intell. Links Theory App. (2013). doi:10.1002/9781118522516.ch9

V. Felipe, J. Espinoza, C. Rojas, J. Rodriguez, M. Rivera, D. Sbárbaro, Multiobjective switching state selector for finite states model predictive control based on fuzzy decision making in a matrix converter. (2013), IEEE Transactions, pp. 1–1

P. Vasant, Hybrid mesh adaptive direct search genetic algorithms and line search approaches for fuzzy optimization problems in production planning. Handb. Optim. 779–799 (2013). doi:10.1007/978-3-642-30504-7_30

Sánchez-Lozano, M. Juan, J. Teruel-Solano, P.L. Soto-Elvira, M.S. García-Cascales, Geographical Information Systems (GIS) and Multi-Criteria Decision Making (MCDM) methods for the evaluation of solar farms locations: case study in south-eastern spain. Renew. Sust. Energ. Rev. **24**, 544–556 (2013)

F. Ribeiro, P. Ferreira, M. Araújo, Evaluating future scenarios for the power generation sector using a Multi-Criteria Decision Analysis (MCDA) tool: The portuguese case. Energy (2013), http://dx.doi.org/10.1016/j.energy.2012.12.036.

Chapter 6
Result and Discussion

Abstract In the present investigation the ideal location for installation of HPP is determined with the help of six different nature based algorithms. The alternative consist three different locations having three different potential of hydro power generations. The algorithms were compared with the real life situations so that an algorithm among them can be selected as the better method for identifying HPP sites accurately but logically satisfying and preventing all the demands and hostilities from the neighborhood people under the requisite geomorphological restrictions. According to the results of comparison with the help of KAPPA method, it was found that BAT followed by Fuzzy algorithms displayed maximum accuracy. The clarity in the decision making of AHP was not satisfactory and the estimated decision of Neuro-genetic model was found to be unreliable. The reason for poor performance of AHP can be attributed to the linear nature of the method whereas for neuro-genetic models the requirement of time and computational infrastructure becomes a constraint. The role of weightage was found to be influential in the accuracy of the output which was evident in the estimated decision from neuro-fuzzy as well as fuzzy logic.

The present investigations utilized the potentiality of popular decision making heuristics in identification of suitable locations for HPP. The nature based algorithms like BAT, Fuzzy Logic, Neural Networks; hybrid algorithms like neuro-genetic and neuro-fuzzy along with AHP was applied to select suitable and optimal alternative from the available options for installation of HPP.

Table 6.1 depicts the results from six different decision making algorithms about the selection of site for HPP. In this regard three sites from three different locations with various geomorphology, climate and Hydrological characteristics was selected where the potential of hydro power is also different. In one place HPP is already installed and in the other two one has no potential of generating hydro power and the last one has moderate (with respect to first location) potential for producing hydro power.

M. Majumder and S. Ghosh, *Decision Making Algorithms for Hydro-Power Plant Location*, SpringerBriefs in Energy, DOI: 10.1007/978-981-4451-63-5_6, © The Author(s) 2013

Table 6.1 Table showing the Output from all the six Decision-making algorithms

Actual Situation	Rank 1	Rank 2	Rank 3
Model Name	River Chenab	River Danube	River Yukon
Neuro-genetic	H(Rank 1)	H(Rank 1)	N(Rank 3)
Neuro-Fuzzy	H(Rank 1)	SH(Rank 2)	N(Rank 3)
Neural Network	VH(Rank 1)	H(Rank 2)	H(Rank 2)
AHP	0.534(Rank 1)	0.173(Rank 3)	0.287(Rank 2)
BAT	69.672 (Rank 1)	54.544 (Rank 2)	42.797(Rank 3)
Fuzzy	2.452 (Rank 1)	1.837 (Rank 2)	1.819 (Rank 3)

Rank with respect to the ratings are also mentioned

According to actual scenario it can be said that Rank 1 will be River Chenab in India (first location). Rank 2 and Rank 3 will be River Danube and Yukon respectively.

Table 6.2 depicts the Hydro-Climatic and Geo-morphological and social status of the three locations included in the study as alternatives for the goal of identifying a suitable location for installation of HPP. Table 6.3 a-r shows the result of agreement test (KAPPA) between the actual and predicted scenarios.

In KAPPA the rank predicted for the locations and Rank assigned to the same was compared, for example, River Chenab is assigned a Rank 1 according to the actual scenario whereas BAT algorithm also ranked the location as 1. Thus both decision making and actual scenario agrees with each other about the potentiality of this location for generating hydro power.

Rank 2 and 3 locations are also compared accordingly.

The results of the KAPPA validation is given at Table 6.3a–r where 6.3.a to 6.3.l depicts the KAPPA results for Neuro-Genetic, Neuro-Fuzzy, Neural Network and AHP algorithms, for Rank 1, 2 and 3 assigned locations. Table 6.3m–o represents the KAPPA coefficient of agreement within the actual and modeled rank by BAT algorithm. Table 6.3p–r shows the agreement results for Fuzzy algorithm.

In case of AHP (Table 6.3j–l), it was observed only 25% agreement for Rank 2 and 3 assigned location when compared with the modeled result for the same locations. But it correctly predicted the rank of Rank 1 location.

In case of neural networks and neurogenetic algorithm, KAPPA for Rank 2 and 3 and Rank 1 and 2 respectively is found to be 40%.

The algorithms BAT, Fuzzy and Neuro-Fuzzy was found to have a KAPPA Coefficient of Agreement for Rank 1, 2 and 3 locations as 100%. That means the rank of all the locations according to the actual scenario is correctly predicted by this three algorithms.

But according to the prominence of decision or the difference of ratings between the alternatives also show how prominently an algorithm can represent a decision. According to Prominence of Decision BAT algorithms was found to have higher difference in rating of the three locations followed by Fuzzy Logic Decision Making process (Table 6.4).

Table 6.2 Table showing the categorical values of the input factors in case of river Chenab, Danube, and Yukon

Factors	Chenab	Danube	Yukon
Average flow	XX	VH	X
Annual probability of average flow	XX	VH	X
Average head	VH	X	XX
Annual probability of average head	VH	X	XX
Water quality IndX	SH	VH	H
Sedimentation	XX	VH	N
Bank erosion	X	VH	N
Turbulence	N	VH	XX
Population concentration	VH	L	SL
Percentage hostile	H	SH	VH
Estimated displacement of population	VH	N	H
Estimated area of affected forest	N	H	VH
Estimated area of affected cropfields	H	N	N
Estimated area of effected waterbodies	H	N	VH
Estimated concentration of effected animal population	N	N	VH
Concentration of endangered/rare animals	N	N	H
Potential of tourism	N	H	VH
Power potential	LX	VH	XX
Utilization potential	LX	VH	XX
Distance to nearest grid	VH	N	N
Distance to nearest consumer	VH	N	N
Slope	LXX	VH	XX
Hydropower potential	Already installed	High but not installed	No potential of installation of hydropower

The present status of hydropower in the considered alternatives are also mentioned

According to the validation results, the neural network model when used singularly, the output from the model erroneously show locations 2 and 3 as Rank 2. The disagreement with the predictions from AHP can depict that linear algorithms like AHP perform less efficiently than nature-based algorithms. Moreover, from the results it can be concluded that the determination of weightage of the deciding variables play an important role in the accuracy of the decisions. All the algorithms, which have utilized the weighted average as output, has predicted in agreement with the actual scenario (Fuzzy, BAT, and Neuro-fuzzy). The reason behind the errors in the applied models output may be attributed to the following:

(1) Requirement of Computational Infrastructures of the hybrid or neural algorithms compared to simple nature-based algorithms like fuzzy or BAT as two different algorithms were required to be computed simultaneously. Even neural

Table 6.3 Table Showing the KAPPA Coefficient of Agreement of Rank 1, 2, and 3 as predicted by the Decision-Making Models

a. Rank 1 as predicted by Neuro-genetic Model			
		Yes	No
Actual	Yes	1	0
	No	1	1
	Yes by both	0.333333333	
	Yes by a	0.333333333	
	Yes by p	0.666666667	
	No by both	0.333333333	
	No by a	0.666666667	
	No by p	0.333333333	
	KAPPA	0.4	

b. Rank 2 as predicted by Neuro-genetic Model			
		Yes	No
Actual	Yes	1	0
	No	1	1
	Yes by both	0.333333333	
	Yes by a	0.333333333	
	Yes by p	0.666666667	
	No by both	0.333333333	
	No by a	0.666666667	
	No by p	0.333333333	
	KAPPA	0.4	

c. Rank 3 as predicted by Neuro-genetic Model			
		Yes	No
Actual	Yes	1	0
	No	0	2
	Yes by both	0.333333333	
	Yes by a	0.333333333	
	Yes by p	0.333333333	
	No by both	0.666666667	
	No by a	0.666666667	
	No by p	0.666666667	
	KAPPA	1	

d. Rank 1 by Neuro-fuzzy Model			
		Yes	No
Actual	Yes	1	0
	No	0	2
	Yes by both	0.333333333	
	Yes by a	0.333333333	

(continued)

Table 6.3 (continued)

d. Rank 1 by Neuro-fuzzy Model			
	Yes by p	0.333333333	
	No by both	0.666666667	
	No by a	0.666666667	
	No by p	0.666666667	
	KAPPA	1	

e. Rank 2 by Neuro-fuzzy Model			
		Yes	No
Actual	Yes	1	0
	No	0	2
	Yes by both	0.333333333	
	Yes by a	0.333333333	
	Yes by p	0.333333333	
	No by both	0.666666667	
	No by a	0.666666667	
	No by p	0.666666667	
	KAPPA	1	

f. Rank 3 by Neuro-fuzzy Model			
		Yes	No
Actual	Yes	1	0
	No	0	2
	Yes by both	0.333333333	
	Yes by a	0.333333333	
	Yes by p	0.333333333	
	No by both	0.666666667	
	No by a	0.666666667	
	No by p	0.666666667	
	KAPPA	1	

g. Rank 1 by Neural Network Model			
		Yes	No
Actual	Yes	1	0
	No	0	2
	Yes by both	0.333333333	
	Yes by a	0.333333333	
	Yes by p	0.333333333	
	No by both	0.666666667	

(continued)

Table 6.3 (continued)

g. Rank 1 by Neural Network Model			
	No by a	0.666666667	
	No by p	0.666666667	
	KAPPA	1	

h. Rank 2 by Neural Network Model		Yes	No
Actual	Yes	1	0
	No	1	1
	Yes by both	0.333333333	
	Yes by a	0.333333333	
	Yes by p	0.666666667	
	No by both	0.333333333	
	No by a	0.666666667	
	No by p	0.333333333	
	KAPPA	0.4	

i. Rank 3 by Neural Network Model		Yes	No
Actual	Yes	1	0
	No	1	1
	Yes by both	0.333333333	
	Yes by a	0.333333333	
	Yes by p	0.666666667	
	No by both	0.333333333	
	No by a	0.666666667	
	No by p	0.333333333	
	KAPPA	0.4	

j. Rank 1 by AHP		Yes	No
Actual	Yes	1	0
	No	0	2
	Yes by both	0.333333	
	Yes by a	0.333333	
	Yes by p	0.333333	
	No by both	0.666667	
	No by a	0.666667	
	No by p	0.666667	
	KAPPA	1	

k. Rank 2 by AHP		Yes	No
Actual	Yes	0	1
	No	1	2

(continued)

Table 6.3 (continued)

k. Rank 2 by AHP

	Yes by both	0
	Yes by a	0.333333
	Yes by p	0.333333
	No by both	0.666667
	No by a	0.666667
	No by p	0.666667
	KAPPA	0.25

l. Rank 3 by AHP

		Yes	No
Actual	Yes	0	1
	No	1	2
	Yes by both	0	
	Yes by a	0.333333	
	Yes by p	0.333333	
	No by both	0.666667	
	No by a	0.666667	
	No by p	0.666667	
	KAPPA	0.25	

m. Rank 1 by BAT algorithm

		Yes	No
Actual	Yes	1	0
	No	0	2
	Yes by both	0.333333333	
	Yes by a	0.333333333	
	Yes by p	0.333333333	
	No by both	0.666666667	
	No by a	0.666666667	
	No by p	0.666666667	
	KAPPA	1	

n. Rank 2 by BAT algorithm

		Yes	No
Actual	Yes	1	0
	No	0	2
	Yes by both	0.333333333	
	Yes by a	0.333333333	
	Yes by p	0.333333333	
	No by both	0.666666667	
	No by a	0.666666667	

(continued)

Table 6.3 (continued)

n. Rank 2 by BAT algorithm			
	No by p	0.666666667	
	KAPPA	1	

o. Rank 3 by BAT algorithm			
		Yes	No
Actual	Yes	1	0
	No	0	2
	Yes by both	0.333333333	
	Yes by a	0.333333333	
	Yes by p	0.333333333	
	No by both	0.666666667	
	No by a	0.666666667	
	No by p	0.666666667	
	KAPPA	1	

p. Rank 1 by fuzzy algorithm			
		Yes	No
Actual	Yes	1	0
	No	0	2
	Yes by both	0.333333	
	Yes by a	0.333333	
	Yes by p	0.333333	
	No by both	0.666667	
	No by a	0.666667	
	No by p	0.666667	
	KAPPA	1	

q. Rank 2 by fuzzy algorithm			
		Yes	No
Actual	Yes	1	0
	No	0	2
	Yes by both	0.333333	
	Yes by a	0.333333	
	Yes by p	0.333333	
	No by both	0.666667	
	No by a	0.666667	
	No by p	0.666667	
	KAPPA	1	

r. Rank 3 by fuzzy algorithm			
		Yes	No
Actual	Yes	1	0
	No	0	2

(continued)

Table 6.3 (continued)

r. Rank 3 by fuzzy algorithm		
	Yes by both	0.333333
	Yes by a	0.333333
	Yes by p	0.333333
	No by both	0.666667
	No by a	0.666667
	No by p	0.666667
	KAPPA	1

Note Here Yes means agreement and No means disagreement between the predicted and actual scenario. Yes by both means both Actual and Predicted decision is same whereas No by both means both Actual and Predicted decisions are not in agreement. Yes and No by a means only actual scenario has matched and not-matched with the rank of the study areas, respectively. Yes and No by p means only predicted scenario has matched and not-matched with the rank of the study areas, respectively. KAPPA displayed the coefficient of agreement as achieved from comparing the actual scenario with the model predicted output by Cohen's KAPPA method where 1 means perfect agreement

network model has correctly decided location 1 as the most suitable option among the considered alternatives but somehow fails to predict the other options in the correct order may be due to the absence of the weightage factor in the output variable (Long et al.1998).

(2) Actuation Error from two different algorithms will be higher in hybrid than the same from single algorithms (Zuperl et al. 2004).

(3) In neural network, the training dataset impacts the rate at which the problem is learned by the model. In the present case, the combinatorial data matrix is relatively large and has innumerous pattern which the model is required to map so that the complexity in the problem can be properly encoded. In case of hybrid algorithms, the weightage are also required to be estimated and the output is actually calculated with the help of this weightage. The complexity of the learning thus get incremented manifold due to these additional step. The result of which may enforce the model to skip many related data patterns which may become eminent due to the shortage of computational power and time allowed for iterations (Psichogios and Ungar 1992).

(4) Another reason can be the categorical nature of the training dataset. The model has to convert the same into its binary equivalent which in turn will take sufficient space and time from simulation infrastructures. Due to the complexity of the conversion, errors may creep in which can jeopardize the way model is interpreting the problem (Mitra et al. 2002).

Again genetic algorithms and neural network both take a large amount of computational infrastructure to complete their objectives. But as fuzzy logic is a lighter algorithm compared to neural network or genetic algorithm neuro-fuzzy has correctly predicted the actual scenario. In case of BAT, the objective of the algorithm was to identify the optimal weightage at which the objective of the decision

Table 6.4 Table Showing the performance rank (based on KAPPA and average difference in the weightage of alternatives), name of algorithm and the justification of this performance of the considered methods to select the better alternative for installation of hydropower plant

Performance rank	Name of algorithm	Reason	Average KAPPA of Rank 1, 2, and 3 alternative (%)	Average of difference within the weightage of alternatives
1	BAT	The main objective of BAT was to amplify the difference of importance of the coherent and noncoherent variables which has incorporated a clear demarcation between influence of coherent and noncoherent variables on the resultant decision	100	23.20
2	Fuzzy logic	Incorporation of the weightage of importance of the deciders into the decision-making variable. Also absence of neural network iterations prevented the necessity of extra computational infrastructure and incorporation of actuation errors	100	13.00
3	Neuro-fuzzy	The incorporation of weightage determined by fuzzy logic have maintained the accuracy of this model but the inclusion of neural network to predict the decision has also introduced the role of actuation error and categorical data set has reduced the level of efficiency of the model	100	Categorical output
4	Neural network	This model does not incorporated the weightage of importance and also the advancement of genetic algorithms to identify ideal topology of the neural network resulted in the erroneous prediction of second and third best alternative in the prediction. The KAPPA for Rank 2 and 3 for this algorithm is 40 %	60	Categorical Output

(continued)

Table 6.4 (continued)

Performance rank	Name of algorithm	Reason	Average KAPPA of Rank 1, 2, and 3 alternative(%)	Average of difference within the weightage of alternatives
5	Neuro-genetic	Both genetic algorithm and neural network are heavy algorithms with requirement of large-scale computational infrastructures. Non-incorporation of weightage of importance also reduced the accuracy of the algorithm yielding a coefficient of agreement of 40 % in case of Rank 1 and Rank 2	60	Categorical Output
6	AHP	AHP is a linear decision-making algorithm to determine weightage of importance of the criteria and alternatives. The decision is taken without considering the original scenario in an objective manner. Rather the subjective nature of assigning importance based on discussion with the experts and market survey has in many cases reduced the accuracy level of the method in predicting the actual scenario. The KAPPA value of 25 % for Rank 2 and Rank 3 alternatives depicts the level of accuracy of the method when compared to other nature-based algorithms like neuro-fuzzy, BAT, fuzzy which have 100 % and even neural networks having a KAPPA of 40 %	50	Categorical Output

making is exclusively prioritized and can be mapped into the output variable which may have contributed to the 100 % agreement with the actual scenario. If predictions from the BAT and fuzzy logic is compared it can be found that the decisions from the former is more pronounce than the latter method of decision making.

Where average difference of rating within the three options for BAT algorithms is 23.2 % the same for Fuzzy is 13 % (approx) (Table 6.4). Although neuro-fuzzy also has clearly delineated the options in three different categories but the requirement of extra infrastructures and other complexities of hybrid algorithms enforced the authors to select BAT as the better algorithm than the six considered in selection of SHP locations.

References

D.L. Long, W.P. Randolph, S.L. Daniel, An introduction to neural network modeling: Merits, limitations, and controversies. Fundamentals of neural network modeling: Neuropsychology and cognitive neuroscience, pp. 3–31 (1998)

S. Mitra, K.P. Sankar, M. Pabitra, Data mining in soft computing framework: a survey. IEEE transac. neural netw. 13(1), 3–14 (2002)

D.C. Psichogios, L.H. Ungar, A hybrid neural network-first principles approach to process modeling. AIChE J. 38(10), 1499–1511 (1992)

U. Zuperl, F. Cus, B. Mursec, T. Ploj, A hybrid analytical-neural network approach to the determination of optimal cutting conditions. J. Mater. Process. Technol. 157, 82–90 (2004)

Chapter 7
Conclusion

Abstract In the present investigation six different nature-based meta-heuristics were used to estimate the decision for selection of locations for installation of hydropower plants. Twenty different factors that can influence the potentiality of hydropower plants were used as criteria and three alternatives have been proposed. The aim of the study was to adjudge the performance efficiency of nature-based algorithms in decision making. That is why all the three alternatives have a known hydroelectricity potential and the decisions were matched with the potential to find the level of accuracy of the algorithm. The BAT was found to be a better performer than the other six models; the reason being that BAT determines weightage at which the difference between coherent and non-coherent variables is the maximum. Algorithms that are linear or where the weightage of importance is not considered have performed poorly. The Fuzzy Logic and BAT, both of which consider weightage for making a decision, has been found to perform satisfactorily. The results from BAT and Fuzzy and according to KAPPA Coefficient of Agreement results make these two algorithms better than the other considered algorithms (neural network, neuro-genetic, neuro-fuzzy and AHP) for prediction of the selection feasibility of a location for installation of HPP. After going through the results it can be concluded that neural network or hybrid algorithms utilized with the same algorithm may have performed poorly due to the requirement of sufficient training data set and computational infrastructures to learn the problem from this given training dataset. Although training dataset is available satisfactorily but limitations in the computational infrastructure may have limited the performance of the said algorithms. In case of hybrid algorithms the same requirement gets doubled. The same study can be repeated with more alternatives and different other algorithms for a more convincing conclusion regarding the potentiality of nature based algorithm in selecting suitable sites for hydropower plants or other related multi criteria decision making problems.

An attempt has been made to select a suitable location for a hydropower plant in such a way that optimal output can be retrieved from minimum but efficient utilization of the available resources. In this regard, the present study utilized six different nature based meta heuristics to estimate the feasibility of a site for installation of a HPP.

The major objective of this investigation was to identify the best nature-based algorithm among the six considered, that can accurately represent the solution to the present problem of site selection for HHP.

The objective was to identify the better algorithm among the six different nature based metaheuristics. Thus, locations where feasibility of HPPs is already interpreted are selected as case study. The three locations for the case study are selected from all over the world in such a way that the decision from the algorithms can be compared and a conclusion about the better algorithm can be interpreted.

The selected locations for this comparative study were:

(1) Upstream of River Chenab in India where HPPs are already installed and operational (one of the largest HPP in India is installed in the Baglihar Dam)
(2) River Danube in Germany where many HPPs are proposed to be installed
(3) Downstream of River Yukon in Alaska where no HPP is ever possible due to the lack of adequate kinetic potential

The six different algorithms that are compared for their efficiency in identification of better alternative for HPPs are:

(1) Neural Networks (One of the most popular nature-based algorithm as discussed in Chap. 4)
(2) Fuzzy Logic(This meta-heuristic is also very popular and widely used in different decision-making studies)
(3) BAT (New meta-heuristic, until now sparsely used as discussed in Chap. 4)
(4) Neuro-genetic Algorithm (To test the performance of Hybrid meta-heuristic)
(5) Neuro-Fuzzy Algorithms (To test the performance of Hybrid meta-heuristic having a weighted output variable)
(6) Analytical Hierarchy Process (To test the performance of one of the oldest Decision-Making Algorithm)

If the real scenario in regard to hydropower generation of the study areas is considered, River Chenab will be the most suitable location and River Yukon will be the least suitable option among the three alternatives considered in the present study.

The neural network models are prepared with the help of combinatorial datasets representing each and every possible combination that can be generated between the input and output variables. The training algorithm Conjugate Gradient Descent was utilized to train the model and identify the optimal weightage by which the network can predict the solution accurately. Here, all the deciders are considered as input and the output was calculated by the ratio of the product of coherent and non-coherent factors. The network topology or the number of hidden layers was identified with the help of trial and error.

In case of Neuro-Fuzzy Algorithms, the output of the neural network discussed earlier was prepared with the help of the ratio of weighted product of coherent and non-coherent deciders, where the weightages were estimated by Fuzzy Logic.

Fuzzy logic was utilized to estimate the weightage of importance for all the deciders. All the fuzzy ratings of importance were converted into their numerical counterparts with the help of scale of importance and then divided by the worst rank achieved by the variable when compared with other variables. The results were complemented and the maximum of the results are collected as the weightage of the variable.

In case of Neuro-Genetic Algorithms, the same neural networks prepared for the same objective were developed, the only difference between the earlier networks and this model being that in the previous case the network topology was predicted by trial and error but in this case the genetic algorithm was utilized to predict the architecture.

In case of AHP the weightage of importance for the deciders as well as the alternatives (based on the deciders) are estimated by comparing with each other.

BAT algorithms, however, are examples of new nature-based meta-heuristic which are slowly but steadily gaining popularity. Here, the food foraging behavior of microbats is utilized in selection of the ideal location. In the present problem, the algorithm was utilized to identify the optimal weightage of coherent and non-coherent at which the difference the two will be maximum. In this way the importance and influence of variables on the decision making objective can be clearly delineated.

Once the models were prepared, the values of the input variables for the three different case studies are incorporated in the algorithms and the decisions were interpreted and compared to find the better alternative among the three considered.

According to the comparison, it was found that BAT accurately and prominently interprets the problem and outputs the ideal decisions that match with the real situations in the case study with respect to other considered decision making algorithms. Moreover, the importance of the weightage given to the input variables to represent their importance in the decision making has endowed its role in accurate decision making at least for the current problem. All the algorithms incorporating this weightage factor have performed better than the other algorithms which have not included the weightage in this output variable. For example, BAT predicted the decision based on the optimal weightage of the deciders and was found to have better accuracy than any other algorithm, followed by Fuzzy which also incorporated only the weightage to identify the better location for installation of hydropower plants. These two algorithms have not utilized the iteration procedures and topology detection phases of neural networks and the hybrid algorithms. Neuro-fuzzy algorithms have also performed more efficiently than the neuro-genetic and neural network algorithms but not as accurately as BAT and fuzzy. The reason may be attributed to the utilization of extra computational resources and actuation errors which are incorporated per iteration of the algorithm due to hybridization. But inclusion of weightage has enhanced its performance more than Neuro-fuzzy and Neural Network algorithms. The neural network model, was also found to yield inaccurate results. The reason behind this poor performance of the model may be attributed to the absence of weightage of importance which was not incorporated in the output variable and using the trial and error method for network topology identification.

Both BAT and fuzzy and also the hybrid neuro-fuzzy algorithm have successfully predicted the suitability of location for HPP and have correctly selected River Chenab as the better alternative over River Danube and Yukon which matches the reality. The hybrid (neuro-genetic), linear (AHP), and algorithms that have not incorporated the importance of the deciders (neural network), however, were found to incorrectly identify the correct alternative.

In case of Fuzzy Logic, the importance of variables was determined with respect to the maximization method. Thus only the importance compared to the dominant variables was reflected in the overall importance of the variable. The BAT algorithm has no such lacunae. That is why the decision from the Fuzzy Logic is not as prominent and as pronounced as BAT algorithm.

AHP on the other hand shows the average importance of the variable and thus the mean importance of the variable is reflected in its overall importance which explained the below par performance of the AHP when compared to BAT algorithm.

The reason behind the supremacy of the BAT algorithm over the other can be attributed to the fact that in BAT no such comparative aggregation of importance was performed and the goal of the algorithm was to clearly differentiate the coherent from the non-coherent variables. The weightage of the variables was determined in such a manner that the difference between the two type of variables of the problem can be clearly incorporated into the decision.

7.1 Limitations

The infrastructure required to train the neural network and its hybrid models was not available for the present investigation. There were over one million datasets which are required to be trained which was impossible with the present infrastructure. That is why batch training was adopted to somewhat rectify this problem.

Another problem of the study was conversion of categorical rating into its numerical counterpart. The models have to convert the categorical data into its binary counterpart and then process the same to retrieve the information available from them. This process requires time and high end computational infrastructures so that the conversion can be performed accurately within the stipulated time frame which was absent to present investigation.

Another drawback of the study is consideration of only three alternatives, but as the aim was to select the better algorithms this shortcoming regarding the number of study areas can be ignored. If required, this same study can be repeated with a larger number of alternatives and with more nature-based as well as nonlinear decision-making algorithms. The time shortage in this project prevented the authors from conducting the study with a greater number of study areas and algorithms.

7.2 Future Scope

This study can be utilized to select suitable locations within a single country, state, or districts. The study can also be repeated with only nonlinear and linear algorithms such as ELECTRE, PROMETHEUS, TOPSIS, etc. The nature- and non-nature-based algorithms can be compared to find whether the former is really more efficient than the latter algorithms.

The validation of decision in this case was performed with the help of KAPPA Coefficient of Agreement. Metrics like Specificity and Sensitivity, Silhouette Index can also be utilized if the correct decision is already known. If the decision is unknown, methods like Voting or Surveying has to be adopted. These methods will enable decision makers to include stakeholder's perceptions about the project.

The present study was an attempt toward logical and scientific decision making for selection of a location which can yield optimal power generation by utilizing minimum resources.

A suitable location for HPP can save a lot of investments as well as energy. Such approaches can highlight the level of logical measure required to be adopted so that minimum cost and energy will be wasted, but maximum profit can be gained. The utilization of this nature-based, pollution-less, infinite, and inexpensive energy source must be optimal due to its nature of uncertainty and our inability in storing the energy for reuse. In light of the population overgrowth and reduction in finite fuel sources, the demand for renewable energy sources like hydropower will rise in the coming decades. Such resources must thus be utilized with greater care and knowledge. Proper decision-making methods that can represent the demand of stakeholders as well as balance the ecological equilibriums only can solve the problem which will become "grave" in the coming future keeping in view the climatic abnormalities and massive urbanization that is taking place worldwide to endanger the available natural resources.